Democratizing RPA with Power Automate Desktop

Boost your productivity by implementing best practices for automating repetitive desktop processes

Peter Krause

BIRMINGHAM—MUMBAI

Democratizing RPA with Power Automate Desktop

Group Product Manager: Alok Dhuri

Publishing Product Manager: Uzma Sheerin

Senior Editor: Nithya Sadanandan

Technical Editor: Maran Fernandes

Copy Editor: Safis Editing

Project Coordinator: Manisha Singh

Proofreader: Safis Editing

Indexer: Tejal Daruwale Soni

Production Designer: Shankar Kalbhor

Business Development Executive: Thilakh Rajavel

Developer Relations Marketing Executive: Deepak Kumar and Rayyan Khan

First published: May 2023

Production reference: 1140423

Published by Packt Publishing Ltd.
Livery Place
35 Livery Street
Birmingham
B3 2PB, UK.

ISBN 978-1-80324-594-2

www.packtpub.com

To my beloved wife, who had to do without me on many evenings,
always supported me in the creation, and had my back.

– Peter Krause

Contributors

About the author

Peter Krause looks back on more than 30 years of IT experience, during which he has experienced the entire spectrum from mainframes to client servers to today's **software as a service (SaaS)** services. In 2003, he witnessed the birth of Microsoft CRM, the foundation of today's Power Platform. From that time on, Microsoft technology has shaped his professional life, where he has held roles in nearly every IT project. Peter is a certified trainer and works at Microsoft Corporation as a solution architect, where he helps enterprise customers successfully build their businesses in the cloud.

I would like to thank everyone who supported me with this book, especially my wife, who helped me with many examples with practical experience and patience and gave up a lot for me during the time of writing.

About the reviewers

Klaudia A. Roveri is a principal solution architect in the Fast Track team in the Business Application Group Microsoft. She has worked for Microsoft for 17 years, performing large-scale implementations of CRM and then Dynamics 365 solutions. Klaudia has experience in complex architecture topics, integration, data migration, and implementation of business processes with a special focus on Power Automate with cloud and RPA areas.

Mehmet Kayir is a solution architect who focuses on Microsoft Dynamics 365 **Customer Engagement** (**CE**) (sales, customer service, marketing, field service), Power Platform, and Azure Stack. He has industry expertise in banking, insurance, and manufacturing. At Microsoft, he supported enterprise customers from scope definition up to design, implementation, integration, and go-live support. Today he works as a lead solution architect for LEADING Business Solution Kayir GmbH and as a partner to train customers (from scale-ups to industry leaders) and help them get up to speed!

Table of Contents

6

Actions for UI Automation   119

7

Automate Your Desktop and Workstation   137

8

Automating Standard Business Applications   165

12

PAD Enterprise Best Practices   255

Index   277

Other Books You May Enjoy   286

Preface

We're all familiar with the situation, in both our private and professional lives, where we have to perform repetitive and boring IT tasks. Microsoft provides Power Automate Desktop, a free tool that can be used to solve both small and large automation scenarios.

In this book, you'll learn everything you need to automate repetitive and monotonous processes with Power Automate Desktop to free up more time for more important things. The book provides insights into the history of the program and its current role in the Microsoft Power Platform. It then explains the concept of user interface automation and how locally installed programs or processes as well as a web browser can be implemented with Power Automate Desktop.

As you progress, you'll learn about the complete feature set of Power Automate Desktop through numerous examples, from basic concepts, such as variables, conditions, and branching, to capabilities with the local desktop, such as file and folder management, to connecting to databases, mainframe computers, and SAP automation.

The final chapters also cover the topic of artificial intelligence and how it can be incorporated into processes, as well as how Power Automate Desktop can be used in large enterprises, where additional topics, such as governance, compliance, scaling, and security, play a role.

Who this book is for

No special IT knowledge is assumed for this book, so an ambitious and process-interested Windows user will be able to find their way around very well, since all the necessary concepts are explained. For the connection of more complex systems, such as SAP, mainframe, or web services, the basic functionality is explained. For their implementation, knowledge in these areas is required.

What this book covers

Chapter 1, *Getting Started with Power Automate Desktop*, introduces you to the concept of Power Automate Desktop with a first example flow.

Chapter 2, *Using Power Automate Desktop and Creating First Flow*, covers installing Power Automate Desktop and creating a first flow by using the built-in recorder.

Chapter 3, *Editing and Debugging UI Flows*, explains the different parts of the application and the fundamental concept of editing and debugging UI flows.

Chapter 4, Basic Structure Elements and Flow Control, introduces using conditionals and loops in UI flows, as well as error handling.

Chapter 5, Variables, UI Elements, and Images, describes what variables are and how they can be used in UI flows, and presents the basic structures for UI automation.

Chapter 6, Actions for UI Automation, continues with the concept introduced in the previous chapter and explains the possibilities of using it to design a process.

Chapter 7, Automate Your Desktop and Workstation, shows how to automate Windows operating systems and services as well as computer peripherals such as the mouse and keyboard.

Chapter 8, Automating Standard Business Applications, represents how Microsoft Office, SAP, and mainframe applications can be automated.

Chapter 9, Leveraging Cloud Services and Power Platform, introduces the concept of Power Automate and desktop flows and how to incorporate IaaS offerings from Microsoft and AWS.

Chapter 10, Leveraging Artificial Intelligence, explains the different capabilities and vendors of AI and how they can be incorporated into UI flows.

Chapter 11, Working with APIs and Services, shows how Power Automate Desktop can also work with different APIs and web services to design an automation process.

Chapter 12, PAD Enterprise Best Practices, explains how Power Automate Desktop can function as part of a larger automation project and what aspects need to be considered.

To get the most out of this book

To work with the content in this book, all that is required is Windows operating system version 10 or 11 and the latest installation of Power Automate Desktop. The installation is described in *Chapter 2*. It also automates various applications, such as a browser (Edge, Chrome, or Firefox), Microsoft Office, and others that may need to be installed.

Some chapters focus on specific applications to be automated, such as SAP or mainframe applications. For this purpose, development or trial licenses were used, the purchase and installation of which are described in the respective chapters.

If you are using the digital version of this book, we advise you to type the code yourself or access the code from the book's GitHub repository (a link is available in the next section). Doing so will help you avoid any potential errors related to the copying and pasting of code.

Download the example code files

You can download the example code files for this book from GitHub at `https://github.com/PacktPublishing/Democratizing-RPA-with-Power-Automate-Desktop`. If there's an update to the code, it will be updated in the GitHub repository.

We also have other code bundles from our rich catalog of books and videos available at `https://github.com/PacktPublishing/`. Check them out!

Download the color images

We also provide a PDF file that has color images of the screenshots and diagrams used in this book. You can download it here: `https://packt.link/nUffQ`.

Conventions used

There are a number of text conventions used throughout this book.

`Code in text`: Indicates code words in text, database table names, folder names, filenames, file extensions, pathnames, dummy URLs, user input, and Twitter handles. Here is an example: "To identify the current file, this gets renamed `work-in-progress.docx`."

A block of code is set as follows:

```
{
    'Brand': 'BMW',
    'Color': 'blue',
    'year of manufacture': '2020',
    'type of vehicle': 'SUV'
}
```

When we wish to draw your attention to a particular part of a code block, the relevant lines or items are set in bold:

```
{
    'Brand': 'BMW',
    'Color': 'blue',
    'year of manufacture': '2020',
    'type of vehicle': 'SUV'
}
```

Bold: Indicates a new term, an important word, or words that you see onscreen. For instance, words in menus or dialog boxes appear in **bold**. Here is an example: "Click on the dropdown for the **All available** entry and change this to **Only the first**."

> **Tips or important notes**
> Appear like this.

Get in touch

Feedback from our readers is always welcome.

General feedback: If you have questions about any aspect of this book, email us at customercare@packtpub.com and mention the book title in the subject of your message.

Errata: Although we have taken every care to ensure the accuracy of our content, mistakes do happen. If you have found a mistake in this book, we would be grateful if you would report this to us. Please visit www.packtpub.com/support/errata and fill in the form.

Piracy: If you come across any illegal copies of our works in any form on the internet, we would be grateful if you would provide us with the location address or website name. Please contact us at copyright@packt.com with a link to the material.

If you are interested in becoming an author: If there is a topic that you have expertise in and you are interested in either writing or contributing to a book, please visit authors.packtpub.com.

Share Your Thoughts

Once you've read *Democratizing RPA with Power Automate Desktop*, we'd love to hear your thoughts! Scan the QR code below to go straight to the Amazon review page for this book and share your feedback.

https://packt.link/r/1803245948

Your review is important to us and the tech community and will help us make sure we're delivering excellent quality content.

Download a free PDF copy of this book

Thanks for purchasing this book!

Do you like to read on the go but are unable to carry your print books everywhere? Is your eBook purchase not compatible with the device of your choice?

Don't worry, now with every Packt book you get a DRM-free PDF version of that book at no cost.

Read anywhere, any place, on any device. Search, copy, and paste code from your favorite technical books directly into your application.

The perks don't stop there, you can get exclusive access to discounts, newsletters, and great free content in your inbox daily

Follow these simple steps to get the benefits:

1. Scan the QR code or visit the link below

https://packt.link/free-ebook/9781803245942

2. Submit your proof of purchase
3. That's it! We'll send your free PDF and other benefits to your email directly

Getting Started with Power Automate Desktop

Power Automate Desktop (**PAD**) is based on principle known as **robotic process automation** (**RPA**). In this chapter, we will look at the essential concepts of PAD and explore the fundamentals that Microsoft leveraged to create the product. In addition, we will look at how the product fits into the ecosystem of Power Platform and what possibilities exist with it.

In this chapter, we will cover the following topics:

- What is PAD and what can it do?
- How it all started – robotic process automation
- Microsoft Power Platform at a glance
- UI flows demystified and their role in Power Platform
- A first example of the incredible potential of PAD and Power Platform

By the end of this chapter, you will have a thorough understanding of the concept of RPA and how PAD fits into the Microsoft automation strategy. The first example will show you the potential of PAD and provide a taste of what you will learn in this book.

What is PAD and what can it do?

To best understand the functionality of PAD, we will go through a few examples in the following subsections where PAD is ideally used.

Example 1 – automatic response letter processing

Imagine you are part of an organizing team at your child's school and an event is to be organized at the school.

Someone in the organization team has already created a form letter in which all the parents of the students are contacted and in which the parents should indicate by ticking whether they agree with the participation of their child. All these cover letters now come back as Word documents and contain the student's names, as well as a ticked box for acceptance or rejection in each case.

You should now generate an Excel list to create an overview of the acceptances and rejections. To do this, you would have to open each Word document and see which student it is and which box was ticked. This whole process is depicted in the following diagram:

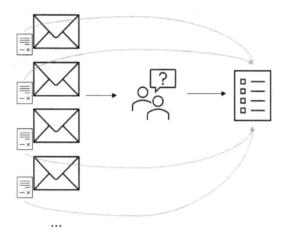

Figure 1.1 – How to generate a report out of email attachments

Depending on the number of students and the information that would need to be looked up, this is a tedious and lengthy task. As you can see from the preceding diagram, this process can be quite boring and time-consuming. On top of that, we would also need to repeat this every day until we reach the reporting deadline.

PAD could help in this situation. It is possible to create a flow that checks all emails belonging to a specific topic and saves the Word document into a specific folder. Another flow can examine all files or documents in that folder, extract the relevant information, and transfer it into an Excel spreadsheet. You would need to create this flow once and it would do the rest of the job automatically.

It would even be possible to make use of **optical character recognition** (**OCR**) if the letters come back as printed output.

Example 2 – integration of a legacy application

Even today, there are still a high number of legacy applications that were created specifically for certain use cases and can't be modernized for a variety of reasons. These reasons include a lack of budget, outdated technology, and a lack of resources, competencies, and time.

The employees are virtually forced to work with this application and very often, such applications cannot access the data or processes programmatically via an **application programming interface (API)** or any other compatible interface.

But what happens when such an application must be integrated into a larger context? For example, there could be a sales platform in which orders and invoices are recorded. However, this data must then also flow into the legacy application to keep the database consistent. Or perhaps, a unique customer number is generated in the legacy application that must be used elsewhere (see *Chapter 7* for an example).

In any case, there must be a way to address this legacy application and integrate it into a larger context – and this is where PAD comes in.

With PAD, we can remotely control any desktop application, record a sequence that is always the same, and then run it again. With PAD, we can also execute this flow remotely or from an external event (the job is entered online in a system), read out the calculation or input results, and play them back as values in the workflow.

This means that an application that cannot be called from the outside and can only be operated manually can nevertheless be automated and integrated. It also implies that in system landscapes, where many applications are involved in a process but have no technical connection, PAD can be used to design a coherent flow and exchange data between these applications.

With its almost 400 actions (at the time of writing; the number of available actions is constantly growing), PAD can cover all user scenarios, beginning with web and desktop applications to databases, files, and folders to email and scripting integration. The following diagram illustrates the capabilities of PAD in terms of connections:

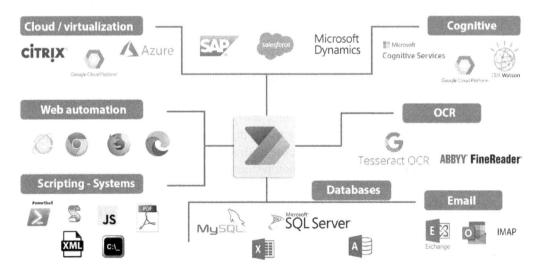

Figure 1.2 – PAD capabilities

Apart from remotely controlling locally installed programs, PAD can also connect to numerous external systems and environments, such as cloud environments from Microsoft or AWS, and incorporate artificial intelligence into scripting and web automation. All these connections can be used in a flow of PAD, allowing cross-system processes to be mapped.

What this illustrates is that PAD can cover a wide range of use cases. However, it also doesn't make sense to use PAD for everything and anything. In the following section, we'll take a look at what criteria a process should fulfill so that the use of PAD makes sense.

What processes should be automated?

As we saw in the previous sections, PAD is very powerful and can be used in private as well as business contexts. Processes and workflows are all around us and we work with them every day, but which process lends itself to automation?

Creating a flow with PAD also involves a certain amount of work, so it's worth taking a closer look to see if it's worth automating a process. Here are some aspects that can be consulted when making a choice:

- Do you understand the process well and is it consistent in how it will be executed? If there are too many variations or unpredictable behavior, it will be very difficult, if not impossible, to automate.

- Is there a sufficient number of repetitions or even a high frequency in which a flow should be executed so that the investment in the work is worthwhile (**return on investment – ROI**)?

- Is the process mainly manual and are there many errors that could be eliminated by automation?

There may be also additional aspects that influence a decision to automate processes, but if you take these questions as a starting point, you are on the right track.

How it all started – robotic process automation

The history of RPA started back in the 1990s–2000s. The need for **user interface** (**UI**), sometimes also **graphical user interface** (**GUI**), software tests became stronger at that time – that is, to be able to automatically click on a software GUI, enter values, and query outputs. For this, the first tools were developed that enabled the first UI and regression tests with the record-and-playback method. Screen scraping technology was also used for this purpose, with which the automation of extracting data from one application was made possible, to use these further in another place. The resulting automatic running process can also be referred to as a **bot** (the short form of **robot**).

Somewhat later in 2005, a company called Softomotive was founded that extended the concept of record-and-playback to enable workflows and data between different Windows applications and thus automate processes even across applications. The product was called WinAutomation and was one of the leading providers of RPA software for a long time, along with other manufacturers.

Softomotive was acquired by Microsoft in 2020 and their product WinAutomation was integrated with its functionality in the Power Automate module, which at that time had very limited functionality to automate desktop-based processes. Nevertheless, Power Platform was already very powerful as an overall concept at this point; we will discuss this in *Chapter 2*.

Microsoft Power Platform at a glance

The actual origins of Power Platform go back very far to the time when there was still talk of Microsoft business solutions. At that time, Microsoft entered the market with a new CRM system, which laid the foundation of today's Power Platform. This ecosystem consists of several subcomponents that can work very closely together but also separately:

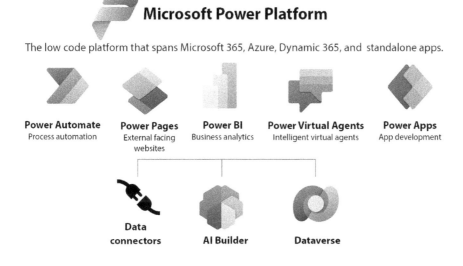

Figure 1.3 – Microsoft Power Platform

These are the components:

- **Power BI**: A tool for visualizing any kind of data in dashboards and reports. Power BI can also capture real-time data from the other components and makes use of the connector capabilities to access data virtually in any data source and combine this with other data to create cross-system dashboards.

- **Power Apps**: This allows users to create different types of apps that can be used for specific business processes. There are different types of apps, which, in turn, can be divided into the form factor (phone/tablet) or the way of creation (canvas versus model-driven). Power Apps also leverages the connector concept to use any data source in an app.

- **Power Automate (ancient Microsoft Flow)**: This is the general term for workflows and business processes that are created within Power Platform. PAD also falls under this category; we will take a closer look at this in the next chapter. But there are also other components here, such as **Business Process Flows (BPFs)**.

- **Power Virtual Agents**: This module allows you to create a so-called chatbot. This can then be embedded into a website or other channel and represents an interactive way for the website visitor to contact the company, ask about the status of an order, and more.

- **Power Pages (ancient PowerApps Portal)**: This module is in preview at the time of writing and comes from the area of portals, which has been around for quite some time. It is a content management system that can be used to create external websites.

The preceding figure also shows three other important building blocks of Power Platform:

- **Data connectors**

  With data connectors, it is possible to access and work with all data sources offered through Power Platform. There are already more than 400 available today and more are being added every day. However, if Microsoft or a third-party manufacturer does not offer a connector, it is also possible to create your own connector and use it in the components offered.

- **AI Builder**

  AI Builder allows you to include artificial intelligence in your apps or workflows. The offered models cover information extraction and text recognition, as well as translation, but also sentiment recognition or classification of customer feedback according to your criteria. Here, too, it is possible to create your own models of the AI, make them available in the platform, and have them used by the app or workflow creators.

- **Dataverse**

  Last but not least, the so-called Dataverse should be mentioned, a very important component of Power Platform. Dataverse is the foundation of Power Platform. All artifacts of the platform are stored and managed here. However, Dataverse is more than just simple data storage. It also contains the security concept, different storage concepts for different requirements, APIs for programmatic access, and much more.

Dynamics 365 workloads store their data models in Dataverse – that is, for a sales module, we find a relational data model with customers and activities, sales opportunities, quotes, and much more. We will see, in an example in *Chapter 9*, how we can use PAD to enrich these modules and how all these components can interact.

From the beginning, the platform has followed the principle that you don't need any programming knowledge to create an app, bot, or workflow. Nevertheless, created components can also be extended by code components, from which the term **low-code/no-code** is derived. Conversely, it does not mean that the platform is not suitable for professional developers. The opposite is true, as Microsoft also invests a lot in supporting professional tools such as **application life cycle management (ALM)** and thus promotes collaboration in so-called fusion teams – that is, citizen and pro developers.

UI flows demystified and their role in Power Platform

As explained in the previous section, Power Automate is a component of Power Platform. However, Power Automate is in itself also an umbrella term for various types of workflows or simply flows that can be created with it. The following types of flows are distinguished:

- **Cloud flows**: These kinds of flows run in the cloud, meaning they are running on the Microsoft Cloud infrastructure within Power Platform. A cloud flow runs on a specific trigger event – for example, *a new contact was created*. It could also run instantly, for example by the press of a button in a Power App, or scheduled – for example, every hour. A cloud flow can be designed in the designer in the browser, and it can incorporate any data or systems that are also available online.

- **Business Process flows**: Dynamics 365 contains a lot of predefined business processes, such as *lead qualification*. This describes the path of a prospect to a real customer, as well as the measures and actions required to achieve this. All these actions are recorded in a BPF and help the user determine what needs to be done next to further qualify the prospect. These flows are used exclusively in so-called model-driven apps and are not relevant in the context of this book. For more effective business processes, it is possible to merge cloud flows and BPFs.

- **Desktop flows or UI flows**: These kinds of flows run on a local machine or workstation that could also be a virtual machine. PAD is used to create and design these kinds of flows, which can access resources such as locally installed applications or the operating system.

> **Note**
> This book focuses on desktop flows and PAD is the tool dedicated to creating these kinds of flows.

PAD contains over 450 so-called actions that can be used to design a flow. These actions and all design capabilities will be explored in the subsequent chapters. In the following subsections, the on-premises data gateway will be explained, as well as the concept of an environment.

Connecting PAD to the cloud

In the previous section, we learned that cloud flows can access data and systems that are online. However, there are scenarios where a cloud flow needs to access a local machine that resides in a home or corporate network (on-premises) – that is, it needs to enable a connection from Microsoft Cloud/ Power Platform to a local machine, as we will see in the following example.

In such scenarios, the issue of security takes on a special focus. For example, companies do not want external access to an internal system, or only under very limited and controlled conditions. The concern is justified, because such scenarios typically require firewalls and ports to be adjusted, which potentially allows an attack from the outside.

Without downplaying the issue on the one hand, but not diving too deeply into the topic of security on the other, the concept of the Power Automate machine runtime is presented here. This provides a secure way to allow access from a Microsoft Cloud system to a local computer.

Within the installation and configuration of the machine runtime on your local computer, you would need to connect to Power Platform and specify the corresponding credentials to register the local machine in the environment. After this, different components within Microsoft Cloud could make use of this connection.

> **Gateways for Desktop flows are deprecated**
>
> A commonly used tool for connecting on-premise data sources is the so-called on-premise data gateway. This tool still exists but is no longer supported for connecting desktop flows. Instead, the Power Automate machine runtime is used to invoke local desktop flows from the cloud. For the use of the machine runtime, certain requirements must be met concerning the version and also the licenses. This will be described in more detail in *Chapter 9*.

Introducing Power Platform environments

Power Platform is the ecosystem in which individual components are created, stored, and executed in environments.

Accordingly, an environment is a container that can wrap around the various components, such as apps and flows, and conversely, each component (app, flow, and so on) always belongs to exactly one environment.

Whether a flow is displayed in the flow list depends on whether the correct environment is selected. We will take this into account in our examples throughout this book.

In this section, we learned that PAD is not only a program for automating local applications, but that it is in a larger context with Microsoft's Power Platform. PAD can be integrated into larger scenarios and can be used in enterprise environments. The next section shows an initial example of how PAD uses Office programs to map an overarching process.

Understanding the potential of PAD by looking at an example with MS Office

Let's take another look at the opening example, in which you were asked to help organize a school event as a parent council. This scenario looks like this:

1. A school event is to take place and the parent council needs to obtain parental consent for the students.

2. Letters have already been sent out to this effect and a Word document has been included. In this document, parents can check off whether or not they consent to their child's participation. Afterward, the parents should send the Word document back to a specific email address. A document for this example has been defined and looks like this:

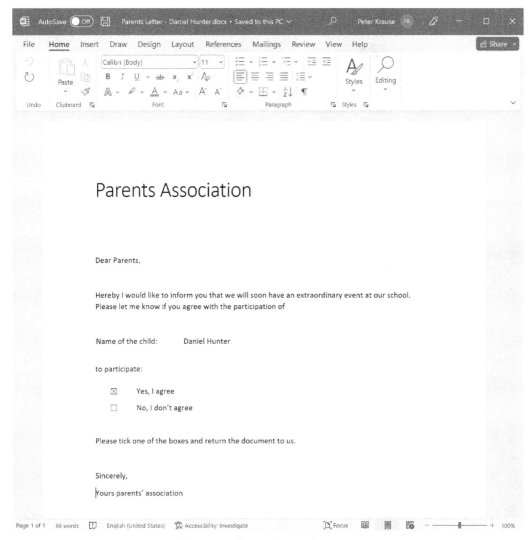

Figure 1.4 – Parents Letter document

3. You now receive a large number of emails, each containing the same document. A summary table with the consents and refusals must be created so that you can do further planning.

This is a good example of a repetitive task that can be nicely automated. The basic procedure is as follows:

Figure 1.5 – Structural view of the flow

The idea here is that you can execute this flow according to your need to collect all responses from parents arriving in the mailbox. Based on this structured flow chart, it is very easy to create a UI flow that does exactly this. Let's take a look at the main building blocks here and how they have been created in PAD. In the subsequent chapters, we will learn how to create a flow like this in detail:

1. *Launch Outlook and save messages to a specific folder.*

    The following screenshot shows the first part of the Desktop flow:

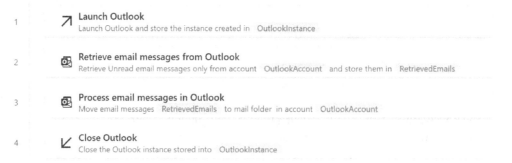

Figure 1.6 – PAD Outlook actions

Each box with a symbol represents an action that can be placed on the design canvas via drag and drop. There are over 450 actions available in PAD grouped into different categories such as **Outlook**, **Files**, and **Folders**. The first four actions are Outlook-related and self-explanatory:

- **Launch Outlook**: Starts the Outlook program on the desktop.

- **Retrieve email messages from Outlook**: Retrieves messages in a specific folder and also allows you to store attachments in messages to be stored in specific folders.

- **Process email messages in Outlook**: This allows you to move messages to some other folder. We will use this to move the messages from the inbox to some subfolder.

- **Close Outlook**: Closes the program.

2. *Open Excel with the tracking workbook.*

The second step is responsible for storing the attached Word documents in a specific folder that is used in the next building block:

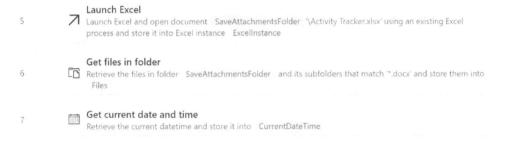

Figure 1.7 – PAD Excel and filesystem actions

Let's look at these actions:

- **Launch Excel**: Opens Excel and also a specific workbook to store the responses.

- **Get Files in a folder**: This action checks the local filesystem and creates a list of files in a specific folder. A filter for files can also be applied. In our case, we only want Word documents with the .docx extension.

- **Get Current date and time**: Here, the flow determines the current date and time, which we can subsequently include in the corresponding entry in Excel.

3. *Handle each document in the attachments folder.*

The following block contains the handling for each Word document. The last sentence contains the for each keyword, which is also used in PAD to iterate through a list of items. In the last building block, this list of items/documents has been created by the **Get files in folder** action and the following block starts with that for each loop, as shown in the following screenshot:

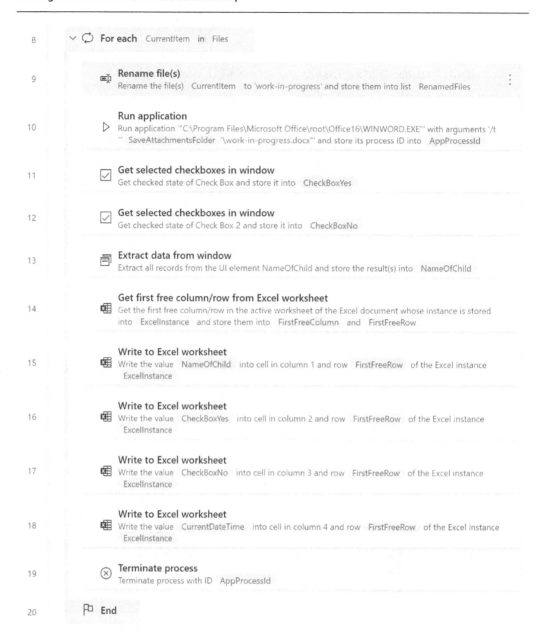

Figure 1.8 – Handling each document in a loop

The actions are as follows:

- **Rename file(s):** Used to rename one or more files. Since we are iterating through the list of Word documents, each iteration contains a different file in the folder. To identify the current file, this gets renamed to `work-in-progress.docx`.

- **Run application**: PAD does not have specific actions for Microsoft Word, but like any other application, this program can also be automated. With this action, we start Microsoft Word on the local machine with a specific document, called `work-in-progress.docx`, which is the current document.

- **Get selected checkboxes in window**: This action can extract `true` (for checked) or `false` (for unchecked) from any window of an application. In our example, we extract the information from the checkboxes of the Word document. We need this for the two checkboxes in the document, which is based on a template that was generated and sent in advance.

- **Extract data from window**: Similar to a checkbox, this action allows us to extract text data from a window. We use this to extract the name of the child in the document.

- **Get first free column/row from Excel workbook**: The workbook has already been opened in one of the actions. This action helps activate the first empty cell so that we can put content in it.

- **Write to Excel worksheet**: This action writes data to the Excel cell specified. We need to perform this action four times to put all the data in.

- **Terminate process**: This closes the Word application so that it can be opened with the next document in the next iteration.

4. *Housekeeping.*

Finally, we need to do some housekeeping, which will be covered by the actions shown in the following screenshot:

Figure 1.9 –Housekeeping PAD actions

In detail, these actions do the following:

- **Delete file(s)**: Microsoft Word creates a hidden file when it opens a document that sometimes isn't deleted and stays in the filesystem. To make sure that we will have valid documents in our folder the next time we run the flow, we just delete the hidden file, if it exists.

- **Save Excel**: Saves our entries in the workbook.

- **Close Excel**: Closes the program.

This UI flow might look quite extensive because of multiple repetitions to store data in Excel, but it is nevertheless easy to create and very powerful. The time saved is enormous and it increases the more that parents need to be contacted and the more messages that need to be processed. In addition, this process could also be extended to using physical letters as input or responding to email messages.

There are, of course, other examples of Desktop Flows and Power Automate that create connections to some external systems or UI flows that get triggered from some cloud system. We will see examples of this throughout this book.

Summary

In this chapter, we started by providing an overview of PAD and how this product has been integrated by Microsoft into their Power Platform ecosystem. We saw that the product fits perfectly into Microsoft's strategy and fills the gap in desktop software integration very well.

In our first example, we learned how different Office programs can be integrated to automate a repetitive process. Instead of handling each message and document separately, PAD saved the documents and extracted the relevant information. Thereby, we were able to save a lot of time by automating this process.

In the next chapter, we will create a flow by taking a small and very useful tool to help us: the recorder.

Further reading

To learn more about the topics that were covered in this chapter, take a look at the following resources:

- *Workflow Automation with Microsoft Power Automate*, by Aaron Guilmette (Packt Pub)
- *Microsoft Power Platform Solution Architect's Handbook*, by Hugo Herrera (Packt Pub)
- Microsoft Power Platform documentation: `https://learn.microsoft.com/en-us/power-platform/`
- About the on-premises data gateway: `https://learn.microsoft.com/en-us/power-platform/admin/wp-onpremises-gateway`
- Power Platform environments overview: `https://learn.microsoft.com/en-us/power-platform/admin/environments-overview`

2

Using Power Automate Desktop and Creating Our First Flow

In the previous chapter, we saw how **Power Automate Desktop** (**PAD**) was born and how powerful the tool can be for automating repetitive tasks. In this chapter, we will take a detailed look into the following:

- What is necessary for installing PAD on a local machine
- How the application is built up
- How to create our first flow by using the built-in recorder

To use PAD effectively, a basic understanding of the application is essential, as we will spend most of our time on the very areas presented here. Also, knowledge of the PAD recorder will be extremely helpful throughout this book because it can also save us a lot of time when creating flows. Now, let's look at how we can get PAD running on our computers.

Understanding the prerequisites for installing PAD

Let's begin by taking a look at what you need to get Power Automate up and running on a local machine. Two things are required to install Power Automate:

- A Windows operating system (Windows 10 or Windows 11 Home edition or above) and some hardware requirements/recommendations
- A Microsoft work or school account to sign in to Power Automate

Since there are specific things to consider, let's take a closer look at each of these topics now.

The Windows operating system

The minimum hardware to run PAD is a 1.00 GHz processor, 1 GB of free storage, and 2 GB of RAM. As always, the more powerful a machine is, the better the applications will run, and this also counts for PAD. If you are running a local application that should be part of automation and that also consumes significant PC resources, these resources should be counted on top.

The Windows Home editions (Windows 10 and Windows 11) have a few restrictions (`https://learn.microsoft.com/en-us/power-automate/desktop-flows/setup`) when working with PAD:

- The first one is that it is not possible to create PAD flows with the Selenium **integrated development environment** (**IDE**). If you want to use this IDE to create PAD flows, a Windows Professional or Enterprise edition will be necessary.

- Selenium is an open source test automation framework for the web. We discussed in the previous chapter that PAD and **Robotic Process Automation** (**RPA**) have their origins in testing frameworks for user interfaces, of which Selenium is an example. Selenium also has an IDE (`https://www.selenium.dev/selenium-ide/`), which you can use to create Power Automate Desktop flows – but only if the underlying operating system is Windows Professional or Enterprise. The second restriction is that a UI flow cannot be triggered from the Microsoft cloud when running on a machine with Windows Home. A PAD flow could potentially be triggered from a cloud flow (Power Automate Flow), and this makes it possible to integrate a local application into a larger context by launching a local workflow triggered by events in the cloud ecosystem.

> **Integration with the Microsoft Cloud**
>
> If you intend to integrate a PAD flow with the Microsoft Cloud, a Windows Professional version or higher is required.

Microsoft account

To be able to access the files for installing PAD, a Microsoft account (formally known as Windows Live ID) is required. There are different types of accounts and all of them are suitable for using PAD:

- **(Personal) Microsoft account**: This is probably the most commonly used type of account in the private area. It is free and provides access to Microsoft's consumer services such as **Outlook. com**, **OneDrive**, and **Office applications**. To sign up, an existing email address can be used or a new one can be created. This account can also be used to sign in to a Windows machine.

- **Work or school account**: This type of account is created by an organization that uses **Azure Active Directory** (**AAD**) as the authentication and authorization platform. Typically, additional services or subscriptions are purchased that offer additional functionality to users.

- **Organization premium account**: This is a special form of work or school account. In this case, the organization is not only managing users via AAD but has also moved certain or all parts of IT to the cloud, such as Exchange, Teams, and Business applications.

With all three types, the full potential of PAD can be unleashed. The organization's premium account has the following capabilities on top:

- Connecting and integrating with cloud flows (see the previous comments on the operating system)

- Collaboration and advanced functionality (such as central Dataverse storage, sharing of flows, as well as centralized management and reporting)

- Additional premium functionality such as access to AI Builder and unattended execution of flows

For the vast majority of use cases, a personal Microsoft account is sufficient. Nevertheless, in the following chapters of this book, we will see scenarios where a premium account is necessary to access specific services.

> **Additional notes on accounts**
>
> To install PAD, a user with administrator privileges on the device is required. A Windows local or domain account with these privileges should be used. To access PAD installation files, however, one of the Microsoft accounts mentioned previously is necessary. The Windows local account can be associated with the Microsoft account in the corresponding Windows account settings.

Now that we know the prerequisites, let's take a look at how to install the software.

Installing PAD

The installation files for PAD are not on the local Windows medium – you must download these from the Microsoft download center. Therefore, an active internet connection and a modern browser are required – Microsoft Edge version 80 or higher or Google Chrome. For this book, I am using a machine with Windows 11 Home edition and a freshly created personal Microsoft account. To install PAD, do the following:

1. Open your browser and navigate to `https://www.office.com`. If you are not already logged in, do so with your Microsoft account.

2. On the right-hand side, you should see a list of familiar icons such as those for **Word**, **Excel**, and more. Click on the last one, called **Apps**. Clicking this icon will make additional apps appear:

Figure 2.1 – Microsoft Office home page with additional apps

3. Click on **Power Automate**. If you don't see this icon, scroll down a bit. Next, you may see a little welcome screen that wants you to choose the right region or country. Please select the appropriate entry and press **Get started**.

4. You will be redirected to the start screen of Power Automate, which is where the installation files are concealed. Click on **My flows** in the left navigation bar:

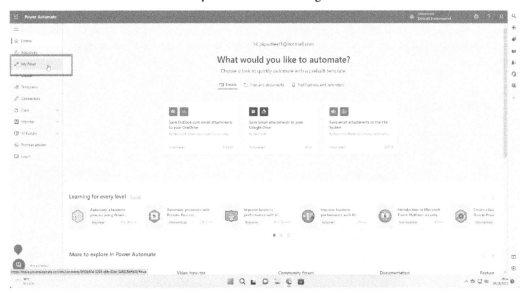

Figure 2.2 – Accessing My flows in Power Automate

5. On the next screen, there is an **Install** button at the top right that reveals the installation files for PAD and the on-premises data gateway. Click on the first entry to start installing PAD:

Figure 2.3 – Installation files for

Another way of installing PAD is via the Microsoft Store:

1. Open the Microsoft Store app and type Power Automate in the **Search** bar.

2. You will be presented with Power Automate as the first entry of the search result. Click the **Install** button to get the tool (I have already installed it, which is why I can open it):

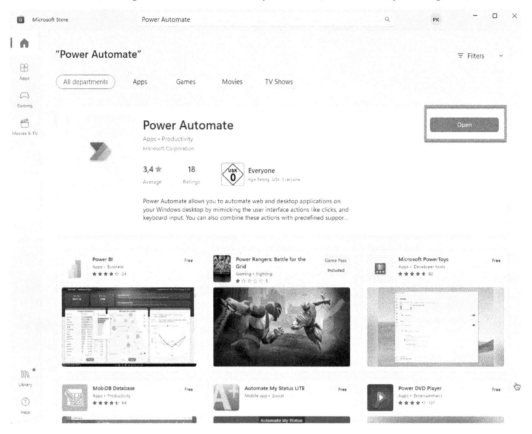

Figure 2.4 – Installing Power Automate from the Microsoft Store

3. Whether it's through the downloaded file or the store, run through the installation process; this is straightforward. You will get a message stating that the installation was successful.

4. After that, click the Windows **Start** button and type Power Automate. The installation procedure does not create any shortcuts, so it is a good idea to pin this program to the **Start** menu or the taskbar:

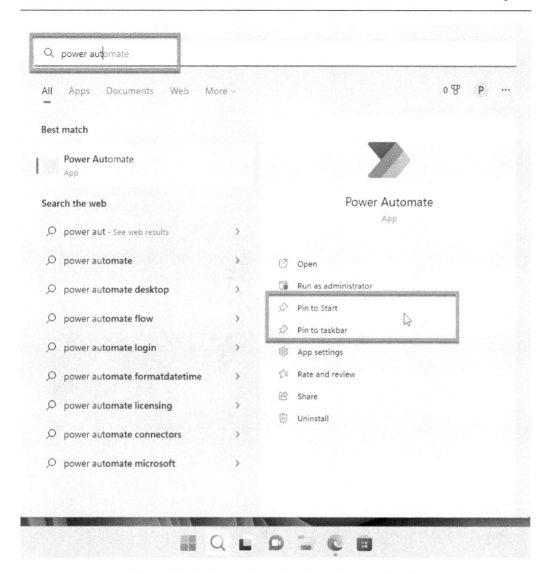

Figure 2.5 – Pinning and launching Power Automate locally

5. Click on the **Power Automate** icon to start the program.

Important note

Please note that the program is called **Power Automate** and not **Power Automate Desktop**. Don't be confused – we are in the right place. The program might find an updated version of the software and apply this automatically.

When the program finally starts, you will also need to log in with your Microsoft account. After this, you should choose a region again and go through the tour of Power Automate, which is also a confirmation that PAD has been successfully installed.

Creating our first flow

Now that we have installed PAD successfully, it's time to create our first simple flow:

1. At this point, the program has started and provides us with a welcome screen. Please note the different tabs, which allow you to switch to your flows or even look at the sample programs from Microsoft, which are worth a look. Starting on the flow container screen, choose + **New flow** either from the top-left menu bar or in the middle of the screen:

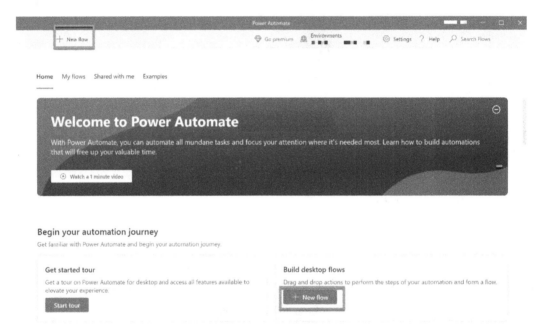

Figure 2.6 – Creating our first flow

2. In the next dialog, you will be asked to name the flow. Although this is optional, it is good practice to name your flow accordingly. Let's choose **Welcome Flow** this time. You can go ahead and click the **Create** button to start the designer:

Build a flow

Figure 2.7 – Flow container screen with the + New flow option

3. This launches the design interface for PAD flows, and this will probably be the place where we will spend most of our time in this book. In the next section, we will learn what all the individual areas mean and what functions they have. But for now, let's create our first flow.

    In our first flow, the idea here is that we want the flow to ask us for our name and then take this to display a nice welcome message. To achieve this, we are going to make use of the **Message boxes** actions. Take a look at the list of entries on the left-hand side and hover over the entry called **Message boxes**; you should see the following:

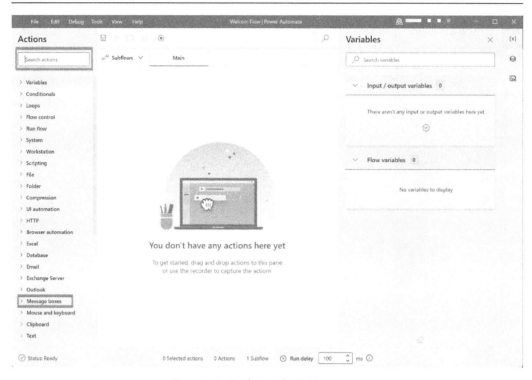

Figure 2.8 – Exploring the Actions pane

4.   Once you get more familiar with the different actions, you can also use the **Search** bar on the top left-hand side to reveal a specific set of actions. Try and click in the search box and enter display as some text. The left list of actions adapts as per the results of the search term:

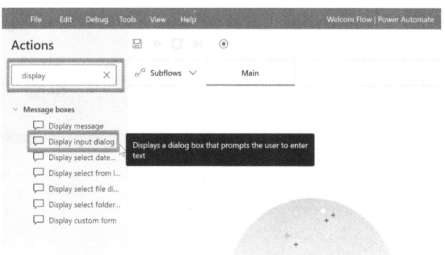

Figure 2.9 – Using the search bar

We are now interested in the **Display input dialog** action, which is the second entry in the list. Remember, you can also use the full list of actions on the left-hand side and expand the **Message boxes** action group:

1. To place the action onto the design canvas, just click and hold the **Display input dialog** entry and drag and drop it into the middle of the window, right under the **Main** tab:

Figure 2.10 – Dragging an action onto the canvas

2. Immediately after you release the mouse button, a dialog will appear. Here, we need to enter the required data:

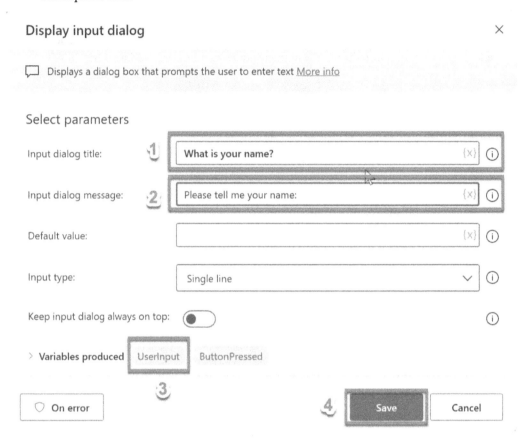

Figure 2.11 – The Display input dialog parameters

We need to provide the following information in this dialog:

I.   Enter a name for the title of the input dialog. This title appears in the header of the message box.

II.   Enter a message for the dialog. In this case, we want to know the name of the person.

III. You don't need to enter anything here. However, the value of the name that is entered by the user must be stored somewhere so that it can be displayed later. This action produces a variable as internal memory for where the input of the user is stored.

IV.   Click on the **Save** button to save and close this dialog.

3.   The design canvas should now contain the first action and look like this:

Figure 2.12 – The first action in the canvas

Now that we have a way to query the name, we want to display that name to the user again. For this, we are going to use the **Display message** action. Click and drag this action and place it under the existing action:

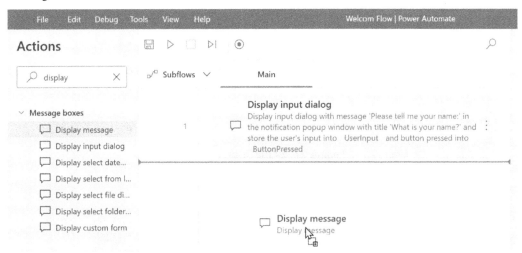

Figure 2.13 – Dragging the Display message action onto the canvas

It's important to note that there is a line indicating where the action will be placed. Later, when we have larger flows with more actions, this will allow us to place an action in the middle of a flow and modify the sequence very easily. Like in our first action, immediately after releasing the mouse, a dialog will appear where we can specify the parameters of this action:

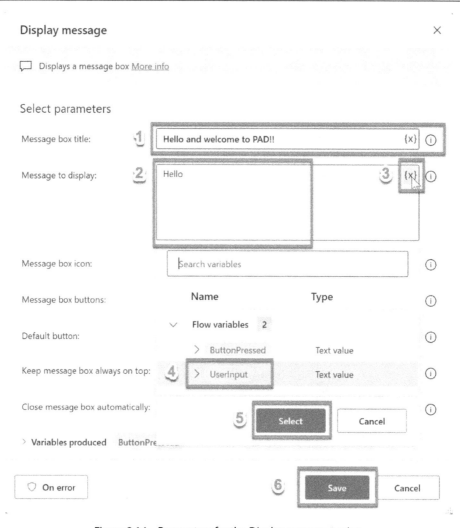

Figure 2.14 – Parameters for the Display message action

Again, we need to enter some information here:

1. This will be the text a user will see in the title of the message box.

2. The message text we want to display.

3. To the right of the text field for the message box, there is a little **X** symbol in brackets. This is a button that can be used to insert dynamic content from variables in the flow. In the previous action, we stored the name of the user in a variable called **UserInput**. When clicking on the **X** button, another dialog will appear at the top that shows us the variables that are available in this flow. Here, we can select **UserInput** and click the **Select** button. The variable will be inserted into the text box for the message. It should look like this now:

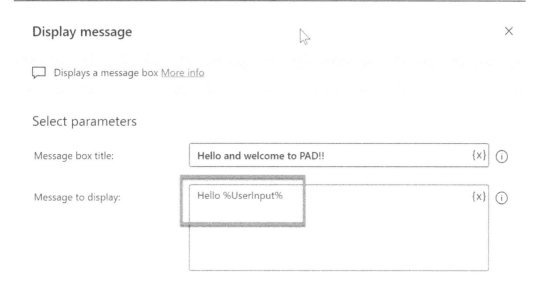

Display message

Displays a message box <u>More info</u>

Select parameters

Message box title:

Hello and welcome to PAD!!   {x} ⓘ

Message to display:

Hello %UserInput%   {x} ⓘ

Figure 2.15 – Using a variable in the message

4. Click the **Save** button to save and close the dialog.

5. Our first UI flow now looks like this:

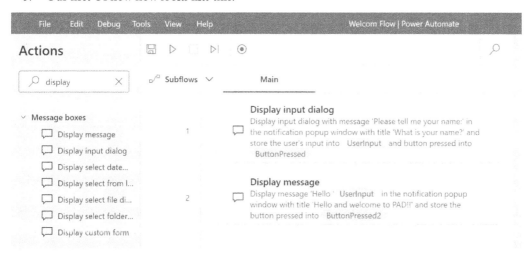

File   Edit   Debug   Tools   View   Help   Welcom Flow | Power Automate

Actions

🔍 display   ✕

Subflows ∨   Main

∨ Message boxes

   ⬜ Display message

   ⬜ Display input dialog

   ⬜ Display select date...

   ⬜ Display select from l...

   ⬜ Display select file di...

   ⬜ Display select folder...

   ⬜ Display custom form

Display input dialog

1   Display input dialog with message 'Please tell me your name:' in the notification popup window with title 'What is your name?' and store the user's input into UserInput and button pressed into ButtonPressed

Display message

2   Display message 'Hello ' UserInput in the notification popup window with title 'Hello and welcome to PAD!!' and store the button pressed into ButtonPressed2

Figure 2.16 – Our first flow completed

6. To test this flow, we need to run it. This can be done by pressing the **Play** button at the top of the action canvas or by going to the **Debug** menu and selecting **Start**. The shortcut for running a flow is *F5*.

As expected, an input dialog appears, asking us for our name, which we can now enter into the text box:

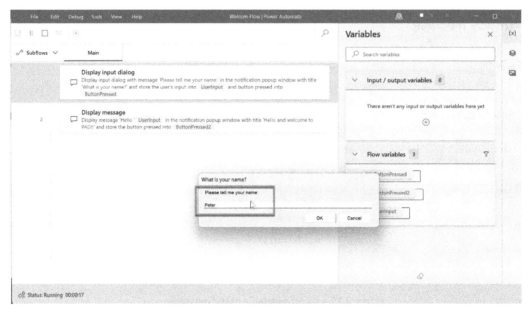

Figure 2.17 – Running our first flow – part 1

Also, notice that in the background, the UI of the program has changed a little bit. The entry with the blue background indicates the current step of the flow. Enter a name and click the **OK** button. The flow will continue with the next step and display a message box with the name you have entered:

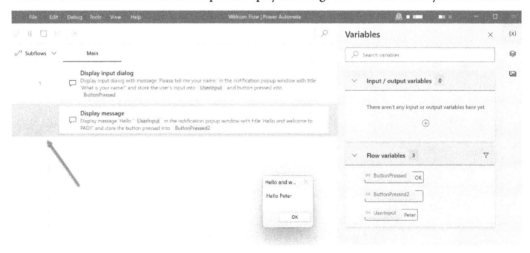

Figure 2.18 – Running our first flow – part 2

Again, the blue background in the application indicates the current step. Click **OK** to close the message box and end the flow. Congratulations – you've created and executed your first UI flow!

Artifacts of the user interface explained

Now that we have explored the basic possibilities of PAD, let's take a detailed look at the application and its possibilities. PAD consists of two windows:

- The container window or Power Automate console
- The Desktop flow designer window

The Power Automate container window

The first window we will see when we start the program is the container window, also known as the Power Automate console. The following screenshot shows this window:

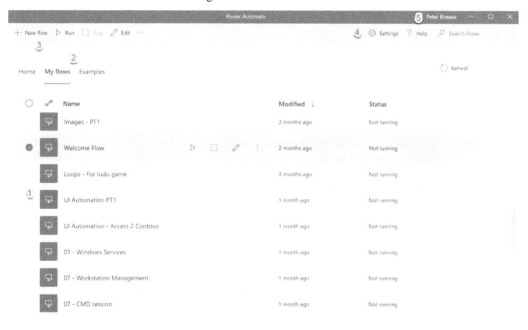

Figure 2.19 – PAD container window

It contains the following elements:

1. **List of UI flows**: If you hover over an item in the list, you will be presented with different additional options, such as the *play* button or the *edit* button. While the *edit* button just opens the designer window (see the following figure), the *start* button runs the flow immediately without opening the designer window first.

You will also get some desktop notifications on the flow's status (has been started, was successful, and so on):

Figure 2.20 – Windows notification on the flow

The three dots to the right of the *edit* button will reveal a context menu where you can also run or edit the flow but also rename, delete, or copy the flow:

Figure 2.21 – Flow list context menu

The **Details** option gives shows the properties of the flow, including its name and description, owner, the date for creation, and the last modification.

2. **The console area switch**: You can also switch from the **My flows** tab to the **Examples** tab. Here, different examples provided by Microsoft are presented, including automation for Excel, desktop, the web, and more. Although the examples do not cover the scope of the program, it may be worthwhile to have a look there to possibly try out one of them.

3. **Create a new flow**: This button launches a dialog where you can enter the name for a new UI flow. Although this is optional, it is reasonable to provide a meaningful name for the flow. There is a **Create** button to confirm the name and launch the designer window with a blank canvas.

4. **PAD settings**: This button launches the **Settings** dialog. Here, we can modify the launch behavior of PAD and also use a hotkey to stop a running flow. In addition, the monitoring and notification settings can be changed (Windows notification, flow monitoring window, or no notification).

5. **Login information**: Clicking on the name in the title bar provides information about the current user and their login information. Here, it is also possible to sign out and log on with another user.

The designer window

Let's take a closer look at the designer window now:

Figure 2.22 – The designer window of PAD

Here, we have the following elements:

1. **The Actions pane**: The left-hand side of the window belongs to the **Actions** pane. Actions represent functionality that a flow is capable of, such as "Open Excel" or "Get files in a folder." These actions are grouped into different categories such as **System** or **Folder** You can see all the actions that belong to a category by expanding the corresponding item. Just click on the arrow icon on the left of the category. Throughout this book, we will look at each of these actions. At the top of the list, there is also a **Search** box. If you enter a search term here, the list of actions gets filtered by this search term. We used this in our first flow.

2. **The flow designer area**: In the middle of the window, there is a canvas that contains all the actions this flow is made of. Actions can be placed there by dragging them from the **Actions** pane. When a flow is started, the actions will run in the sequence in which they are listed here. This action list can be rearranged by just clicking and dragging an existing entry and putting it somewhere else. There will be a horizontal line displayed showing where an action can be placed. We saw this in our previous example. You can also double-click on an action to bring up the settings dialog for that action and modify the entries. In this area, there is also an option to look at the subflows. A subflow is just another UI flow that encapsulates a specific set of actions. By using subflows, it is possible to structure a large flow into different pieces. A subflow always belongs to a regular flow and is not shared between other flows.

3. **The flow control bar**: The little bar at the top of the action list allows you to save, start, stop, and record a flow. There is also an option to start the flow and go action by action, which is also called **debugging**. We will take a closer look at this, as well as the recorder, in the next section.

4. **The Variables viewer pane**: On the right-hand side, there's an area that displays the variables of the flow. Variables are an important aspect of a flow for storing information during runtime. Variables will be covered in *Chapter 5*.

5. **The input elements switch**: Please also notice the two additional little icons below the variables icon, {**x**}. These can be used to switch to the area containing UI elements and images, which will also be covered later. You can hide this area by clicking on the icon.

6. **Action and execution summary**: At the bottom, there is a little bar that summarizes the current flow. It is also possible to set a specific delay for the flow to run – in this example, 100 milliseconds.

    This is the applications menu bar. In addition to the typical functionality for saving and editing, the debug functionality is also available here once again, in which breakpoints can also be set. This allows a flow to be executed up to a certain point so that we can check the settings and variables in the event of an error:

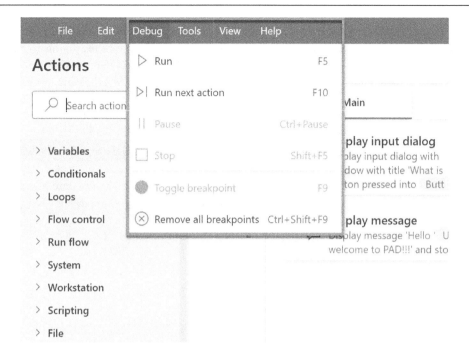

Figure 2.23 – The Debug menu

This designer window will also show up in another area at the bottom of the screen when a flow runs into an error. There will be detailed information about where the flow stopped and why. Now that we have a few more details, we'll use the recorder to create another flow in the next section.

Making use of the recorder

As you saw in the previous chapter, the PAD designer window allows us not only to drag and drop actions onto the design canvas but also allows us to record a UI flow. This concept is comparable to a screen recorder that records every step, every mouse click, and keyboard stroke to play back this sequence the same way later.

We want to use this functionality once to create an overview of all book titles of the PacktPub publisher so that we can use this information elsewhere – for example, in a database. Unfortunately, we don't have direct access to PacktPub's underlying source, but we can get the book titles via their website at `https://subscription.packtpub.com/search`. However, we don't want to highlight every single title on the website and then copy and paste them into another tool. That would take a very long time and also be quite boring. PAD can do this job for us.

In the following example, we will use a concept known as web scraping – that is, structurally extracting information from web pages. Note this also works with numerous other websites, so this technique can be used universally. However, the prerequisite for using this method is that we must install and activate a small extension for the browser we're using so that PAD can access the elements of a web page. So, let's get started:

1. Create a new flow and give it a meaningful name, such as **Web Scraping PacktPub**.

2. Locate the **Browser automation** actions group in the actions pane and expand it. You will see a **Launch new...** action for each of the browsers from Google, Firefox, and Microsoft Edge. I have Microsoft Edge installed on my machine, so I will click and drag this action onto the canvas:

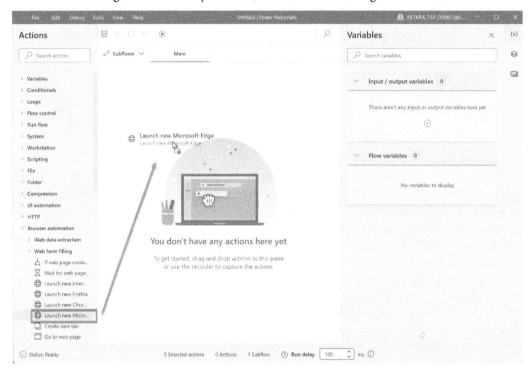

Figure 2.24 – Dragging the browser action onto the canvas

3. At this point, you might be requested to install the browser extension for PAD:

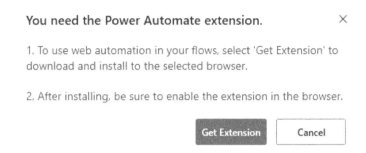

Figure 2.25 – Browser extension installation dialog

4. This will only happen once. Get the extension and make sure that it is activated.

    In the dialog for this action, enter the data accordingly and make sure that you use `https://subscription.packtpub.com/search` as the initial URL. Click on the **Save** button:

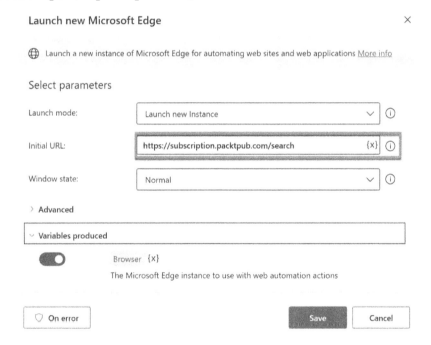

Figure 2.26 – Settings for the browser launch action

You should now have one action in the canvas that opens the website mentioned previously in a new browser instance. Test this flow now by pressing the **Run** button above the design canvas and leave the browser window open for now.

As mentioned previously, we can always add more functionalities to the flow. For this, we will use the recorder. With the browser window open, switch back to the design canvas/workspace window and click the **Recorder** button:

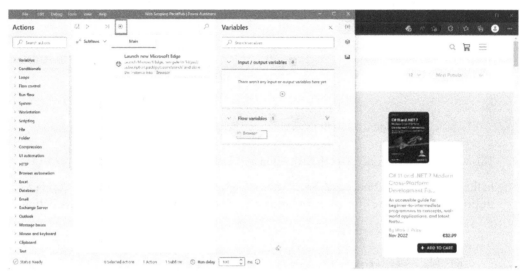

Figure 2.27 – The recorder button

At this point, the PAD designer will disappear, and a smaller recorder window will appear instead. It should look like this:

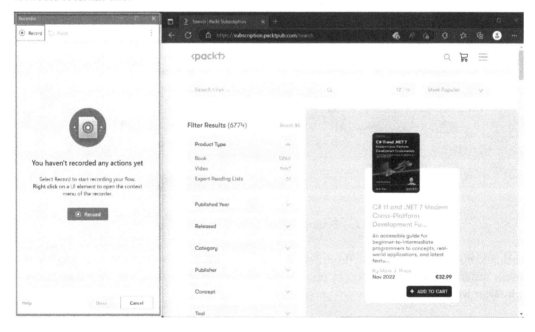

Figure 2.28 – The recorder window

Now, the funny part begins. Start the recording by either clicking the button in the middle of the screen or at the top. The recorder window will change and display an empty area under **Recorded actions**. Use your mouse and hover over different elements on the website, such as the book picture. You will see that PAD tries to identify the different elements on the website and frames them accordingly. In the following screenshot, I hovered over the book image:

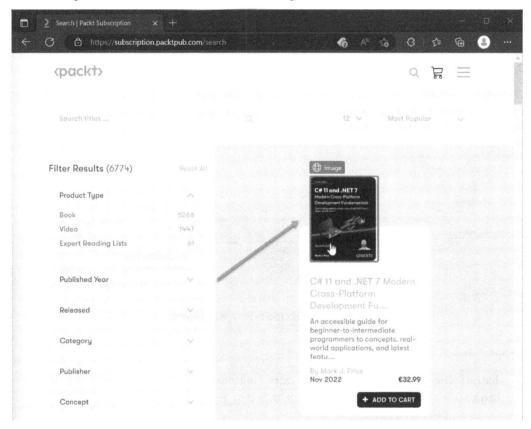

Figure 2.29 – Detecting web elements

For the recorder to recognize the web page elements correctly, it sometimes takes a bit of dexterity. It also helps to move the mouse out of the desired area and back in again to get the right mark. Once the correct element has been highlighted, click the right mouse button to be able to select the information we want to extract from this element. A context menu for this element will appear so that we can select different entries:

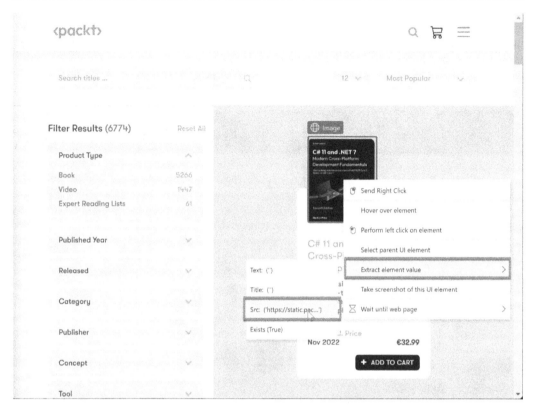

Figure 2.30 – Context menu for web elements

Let's take a look at this in more detail:

- **Extract element value**: This allows us to extract information from this element. A web element can always have different properties, such as **Text** or **Title**.

- **Wait until web page**: This allows us to wait for specific events on the page, such as when an element contains a text or gets visible.

- **Take screenshot of this UI element**: This takes a screenshot and stores it in PAD flow.

In our case, we may want to extract the information of the image source, so we will choose **Extract element value | Src**. The recorder window should look like this now:

Figure 2.31 – A recorded action in the recorder

> **Important note – remove additional actions**
>
> You might have other actions that have been recorded by the recorder, depending on where you clicked and what keys you might have pressed. For our example, there must be only one action in the recorder. Use the **Delete** button (trashcan symbol) to remove any additional actions.

Now, repeat the same process for the title of the book. Use your mouse to hover over the elements and the title and click the right mouse button:

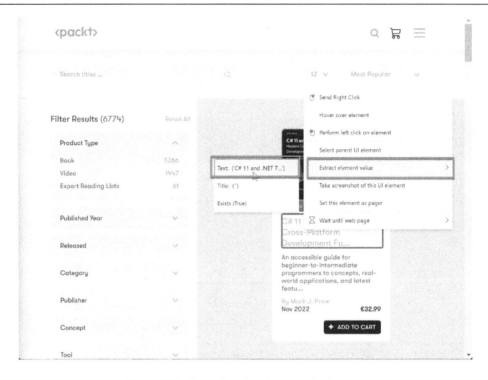

Figure 2.32 – Extracting data from a web element

This time, we want to extract the title of the element. With that, we have captured two elements on the web page, which are represented in the recorder window like this:

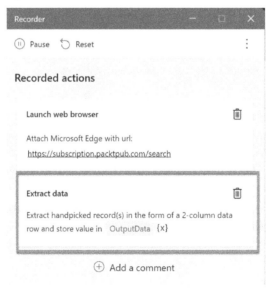

Figure 2.33 – Extracted data in the recorder

There is an important difference here: PAD already recognized that we captured two pieces of information from one record on the web page. Our next step will be to go to the next element, and this is where some magic starts. If you repeat the first step and mark the next book image and extract the **Src** information again, PAD will recognize this pattern and assume that you want to do the same as with the first element. We can see this on the web page because PAD has marked the corresponding elements with a green dotted line:

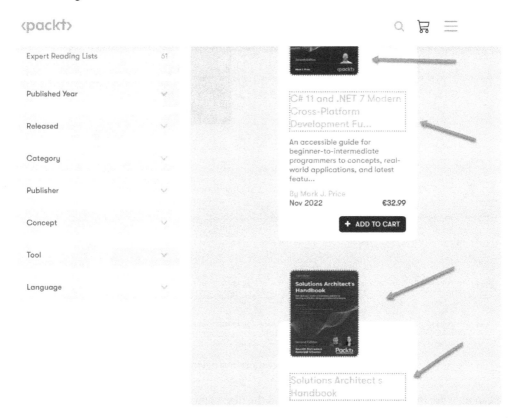

Figure 2.34 – Pattern recognition of the recorder

Pattern recognition

Please notice that it sometimes takes capturing more than two structures before the recorder recognizes the pattern. So, please proceed to the next element if PAD does not recognize the pattern immediately. Recognition also depends on the underlying structure of the HTML. After the third or fourth element at the latest, the pattern will be recognized.

This means that PAD will capture all the elements on this page now. But what about the next page of books? Let's scroll down so that we can see the pagination control on that website:

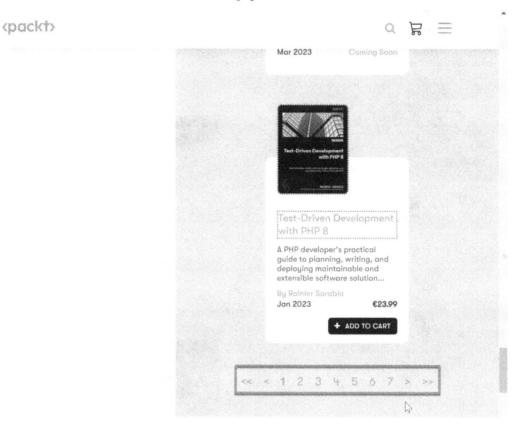

Figure 2.35 – Incorporating navigation

We can now use the single arrow button to load the next page. Right-click on the element shown in the following screenshot and choose **Set this element as pager**:

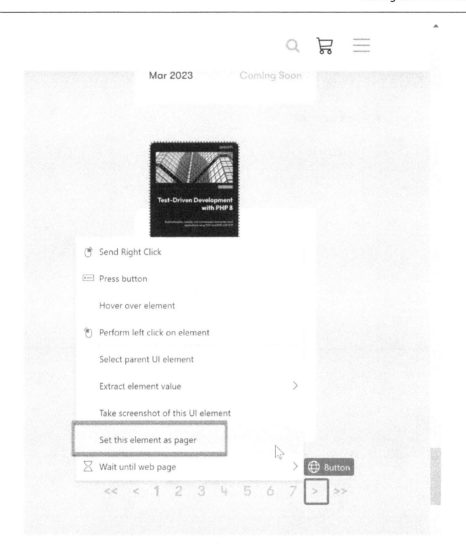

Figure 2.36 – Using a pager in a recording

The element now gets a little blue dotted border, indicating the pager functionality. Looking at the recorder window, just one element should be seen here:

Figure 2.37 – Recorded web scraping actions

If other elements have been captured by mistake, just delete them. You can now stop the recording by clicking the **Done** button. Now, the recorded actions get transferred to the main PAD design window:

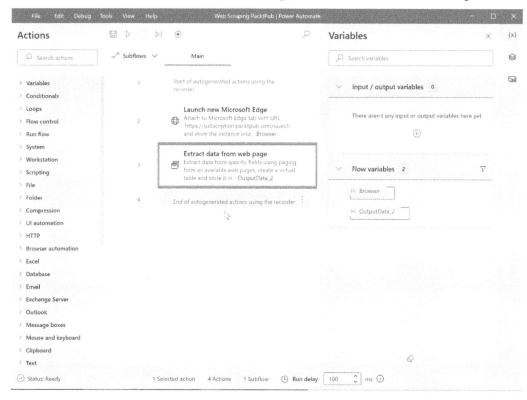

Figure 2.38 – Transferring recorded actions to the designer

There are also comments in the list of actions describing the start and end of the generated actions by the recorder. Let's investigate the **Extract data from web page** action a bit more by double-clicking it:

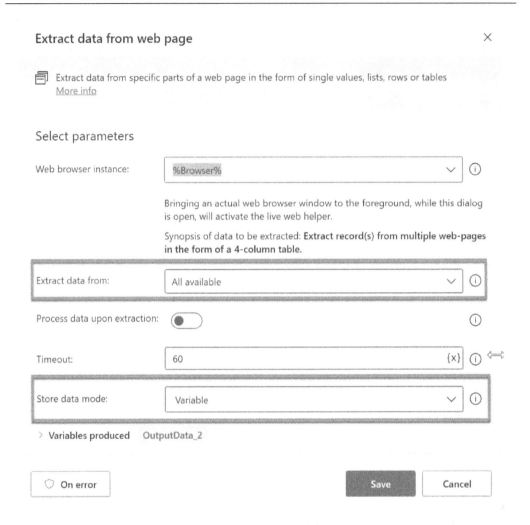

Extract data from web page   ✕

📑 Extract data from specific parts of a web page in the form of single values, lists, rows or tables
     More info

Select parameters

Web browser instance: %Browser%   ⌄   ⓘ

Bringing an actual web browser window to the foreground, while this dialog is open, will activate the live web helper.

Synopsis of data to be extracted: **Extract record(s) from multiple web-pages in the form of a 4-column table.**

Extract data from: All available   ⌄   ⓘ

Process data upon extraction: ⬤──   ⓘ

Timeout: 60   {x} ⓘ ⇔

Store data mode: Variable   ⌄   ⓘ

> Variables produced   OutputData_2

🛡 On error   **Save**   Cancel

Figure 2.39 – Revealing the web data extraction

The interesting parts here are **Extract data from** and **Store data mode**. After the browser has been loaded by our first action in the flow, this action extracts all the data from the web page unless we don't tell it otherwise. While on some occasions this might be useful, for our test, we only want to extract the first 10 pages. Click on the dropdown for the **All available** entry and change this to **Only the first**. Now, another text box will appear, where we can enter the number of web pages to process.

The last setting for this example is **Store data mode**. So far, the data that is captured is stored in a variable (`DataFromWebPage`, in my example). But of course, there are other ways to work with this data now. For example, you could take this variable and use it as input for another action after the web data has been extracted, or you could choose to store this in Excel. Therefore, click the dropdown for **Store data mode** and change the value from **Variable** to **Excel spreadsheet**:

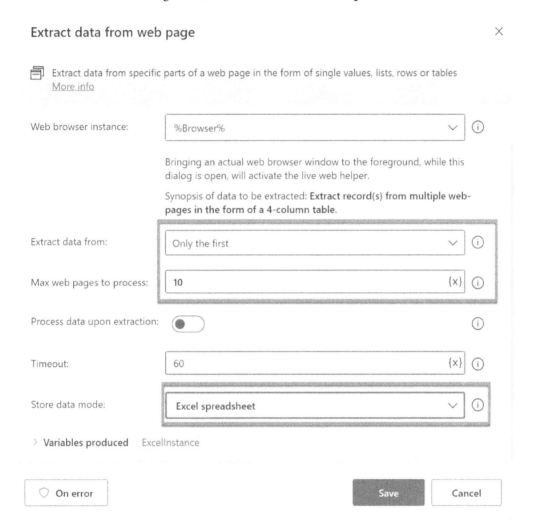

Figure 2.40 – Modifying the web data extraction parameters

Now, the web data extraction will only run for the first 10 pages, take the data, launch Excel with a new spreadsheet, and paste the captured data in there. Save the dialog now and let the show begin. On my machine, it took 45 seconds to extract 120 records and paste them into Excel.

Summary

Installing PAD is straightforward, just like installing any other Microsoft tool. Once you have signed up for a Microsoft account, you can log in to the program and start creating your UI flows right away.

It is recommended to try things out and use the tool even to automate simple tasks. This will make you more familiar with the program, as I explained in this chapter. The recorder can save you a lot of work and it can be used for new or existing flows. Especially if web scraping is a task, this tool can be very handy.

In the next chapter, we will continue our journey and look at the editing and debugging capabilities of PAD.

Further reading

To learn more about the topics that were covered in this chapter, take a look at the following resources:

- *Install Power Automate Browser extensions*: https://learn.microsoft.com/en-us/power-automate/desktop-flows/install-browser-extensions
- *Record desktop flows*: https://learn.microsoft.com/en-us/power-automate/desktop-flows/recording-flow

3
Editing and Debugging UI Flows

In the previous chapters, we learned how to create and execute a UI flow using the designer window and the recorder. In this chapter, we will look at what additional editing options are available and how we can stop a UI flow and check for data or errors. We will also learn a few tricks and tips that will make it easier to create and test larger flows.

We will cover the following main topics in this chapter:

- Details of the designer window/workspace
- The parameters dialog
- Global menu options
- Using breakpoints
- Working with subflows
- Strategies for effective flow creation

By the end of this chapter, you will have a solid understanding of the flow designer and how to use it effectively to create, edit, and structure flows.

Details of the designer window/workspace

A UI flow and its content are created by dragging actions from the **Actions** pane into the workspace. The workspace, by the end, contains a list of actions that run in sequence from top to bottom. This sequence can easily be changed; we just have to click and drag the action to another place up or down. In the following screenshot, we will use the example from the first chapter, where we saw the automation of different Office applications:

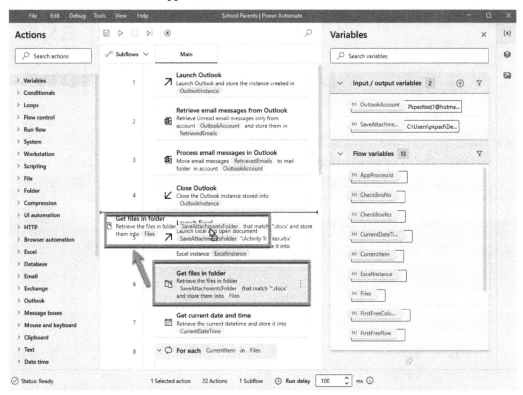

Figure 3.1 – Dragging an action into the workspace

The bold horizontal line in the workspace shows us after which action the new position will be. In the preceding example, we are dragging **Get files in folder** up and placing it after **Close Outlook**.

Once the actions have been set in the right order, we can use the kebab menu (the three vertical dots) to access additional options. The same happens when you use the left mouse button.

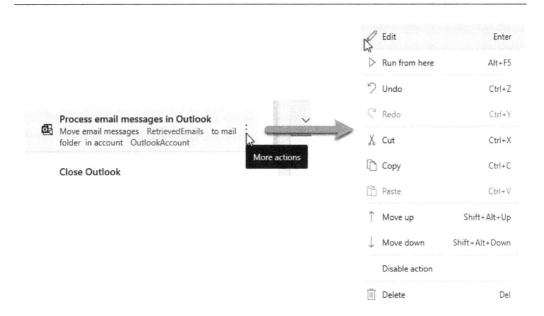

Figure 3.2 – Action kebab menu options

We have the following options:

- **Edit**: This pops up the **Parameter** dialog for the specific action. This dialog can change depending on the action we are using. A **Launch Outlook** action has other settings than **Get files in folder**.

- **Run from here**: If we use this command (or *Alt+F5*), the flow will start to execute from this action onward. All previous actions will not be executed. If there are other actions before, which are necessary for our selected action to be executed, this might lead to an error. But nevertheless, it could also be a handy function if the flow ran into an error previously, and you just want to start right there.

- **Undo** / **Redo** / **Cut** / **Copy** / **Paste**: These are well-known editing functions from other programs, such as Office.

- **Move up** / **Move down**: Instead of clicking and dragging an action, we could use these commands or use shortcuts.

- **Disable action**: This is also a very useful feature when designing a flow. It allows you to disable an action so that it will not be executed. For example, you might want to do some housekeeping at the end of a flow and delete files or close a program. But, if you are still in the middle of the design, you could disable these actions so that you can concentrate more on the design. This is also helpful when debugging a flow (which we will do in the next section). But be careful not to disable steps that are part of loops or conditionals. These actions encapsulate one or more other actions with a beginning and an ending. The following picture shows an if conditional action and another if process action. The corresponding end statement for the latter condition is disabled, which leads to an error.

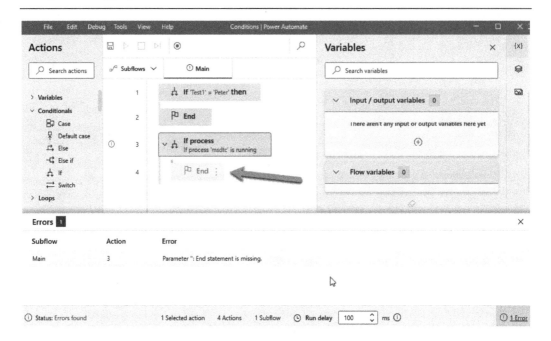

Figure 3.3 – Disabling an action and error message

- **Delete**: Instead of disabling an action, it could, of course, be deleted, which removes the action and all the parameter settings from the flow.

The parameters dialog

As soon as we drop an action from the **Actions** pane onto the workspace, a dialog pops up requesting us to provide information – the parameters dialog. The following screenshot shows the parameters dialog of the **Get files in folder** action:

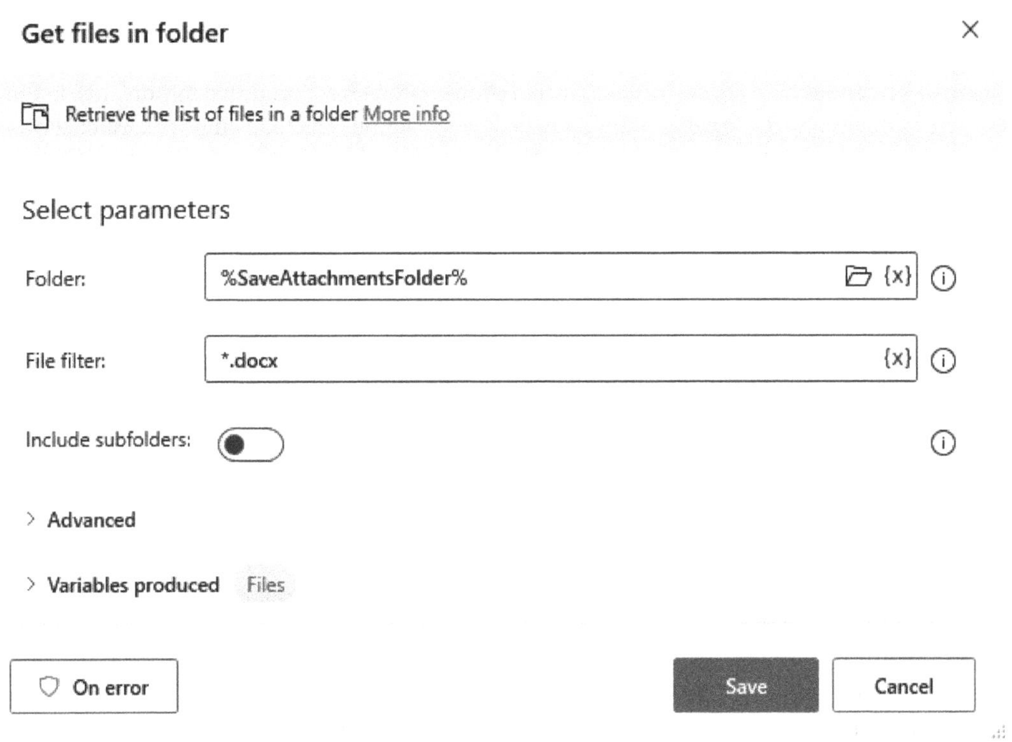

Get files in folder   ✕

🗏 Retrieve the list of files in a folder <u>More info</u>

Select parameters

Folder: %SaveAttachmentsFolder%   🗁 {x} ⓘ

File filter: *.docx   {x} ⓘ

Include subfolders: ⬤───   ⓘ

⟩ Advanced

⟩ Variables produced   Files

🛡 On error   Save   Cancel

Figure 3.4 – The parameters dialog

This dialog shows the specific parameters for the chosen action. Since almost every action is different, the corresponding parameters dialog will also show different parameters. There are indeed actions that have similar parameters, for example, **Copy file(s)** and **Move file(s)**, but these are more or less the exceptions.

The parameters dialog lets you enter all the necessary information that the action needs to be executed. This information can be text or numbers, the selection of a drop-down menu, a toggle switch such as the one in the previous screenshot (**Include subfolders**), and sometimes even more complex information.

Sometimes, the parameters are also divided into the most used and advanced ones. These are hidden when the dialog first appears but can be revealed by expanding the **Advanced** area. We can also see this area in the preceding screenshot.

> **Important Note**
>
> The parameters dialog does not show the mandatory fields required for the action to work. The only way to find out whether a parameter is mandatory is to leave it blank and save the action. Missing information will be displayed as an error in a message window at the bottom of the window. The UI flow will not run until these errors have been corrected.

Actions can also produce or capture information that could be used in some of the next steps in the flow. For example, in *Figure 3.4*, the **Get files in folder** action creates a list of files in the given folder. To make this information available for subsequent actions, variables are produced to store this. In the preceding example, a variable called **Files** is created by the action. The following screenshot shows that this section of the dialog can also be expanded:

Figure 3.5 – Expanded variables area

When expanded, it is possible to rename the variable accordingly. Here, we renamed the variable to MySpecialVariable. There are naming conventions which we will discuss in the next chapter in more detail. If the data of a variable will not be used in the flow, it is possible to disable the creation of the variable by pressing the corresponding toggle switch.

After an action has been provided with all the required information, it can be saved. Variables that are produced by the action will then be displayed in the **Flow variables** area on the right side. And now, these variables are also available to be used as input for another action. To use a variable as input, we don't hardcode information into the parameters input box, but instead, use the **{x}** symbol on the right. Check out *Chapter 5* for more information on variables and to get all the details.

Now that we know the details of the designer window, let's take a look at the additional functionality provided by the global menu options.

Global menu options

We already touched upon the menu bar in the previous chapter. These functions can be very valuable when we are editing a flow. Let's take a look at the most important menu options:

- **File | Save / Save As…**: Save a flow or create a copy by using the **Save As…** option and provide a new name for the flow.

- **Edit | Undo / Redo**: These are well-known options that allow you to undo or redo any action in the designer. This is also very useful if you accidentally delete an action.

- **Edit | Select all / Clear selection / Invert selection**: One or more actions in the workspace can be *selected*. If you click on one action, it becomes selected. Holding down the shift key and selecting another action will result in all the actions in between will also be selected. This group of actions can now be deleted, activated or deactivated, or even moved together to some other position in the workspace. These menu options deal with this selection.

- **Edit | Go to line**: Sometimes it is nice to be able to jump to a specific line immediately, for example, when working with large flows and many actions or in case of an error because the error message also contains the line number.

- **Debug | Toggle breakpoint / Remove all breakpoints**: When an action is selected, this menu option will place a breakpoint on that line. Another way to do this is by clicking directly on the left side of the line number. We will discuss breakpoints more in the next section.

All menu items also have keyboard shortcuts. By internalizing those, a very fast and effective working style in the workspace can be achieved.

Using breakpoints

Breakpoints are a common concept in programming languages. A breakpoint is like a stop sign for the flow. It stops the execution and gives you a chance to inspect the content of variables or some other system. You can create a breakpoint through one of the menu options mentioned previously. The following screenshot shows a breakpoint in line **3**:

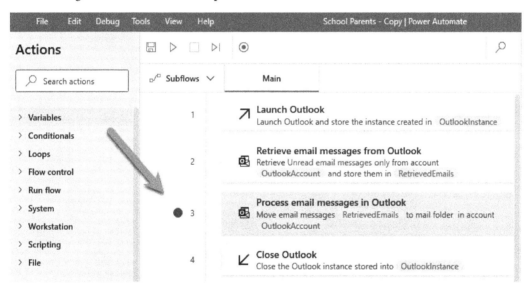

Figure 3.6 – A breakpoint in the workspace

When you run this flow, it will stop before the action with the breakpoint, meaning that this action has not yet been executed and the flow is paused. You can do one of the following actions:

- Continue to run the flow by pressing the *play* button at the top or *F5*

- Run only the next step by pressing the button to the left of the recorder button or *F10*

- Stop the flow with the *stop* button or *shift + F5*

If the flow is stopped, all execution results of the actions before that breakpoint will be preserved. As seen in *Figure 3.6*, the breakpoint is after the **Launch Outlook** action. If we stop the flow execution now, Outlook would still be open. This behavior is also helpful when following a phased approach for flow design (see the next section). Breakpoints can also be very useful in controlling the values of the variables. We will make use of breakpoints in the next section.

Another option to pause a flow is to use the *pause* button, which becomes visible when the flow is running. In fact, the *play* button changes to a *pause* button, as you can see in the following screenshot:

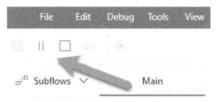

Figure 3.7 – The pause button

This also allows you to pause the flow anywhere without using a breakpoint.

Working with subflows

Within each flow, we can also create subflows. This is a concept to structure a larger UI flow into different pieces, which in turn can then be called and referenced in the main flow:

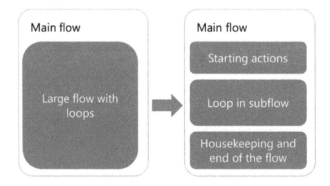

Figure 3.8 – Deconstructing a flow

The main flow containing the overall logic of a task then becomes much more readable and easier to understand. All variables are available in the main flow and can be changed, no passing parameters must be specified. It is even possible to call different **subflow** from within a subflow if you want to structure the work even further. The following figure shows the main flow from our previous example and two subflows:

Main Flow

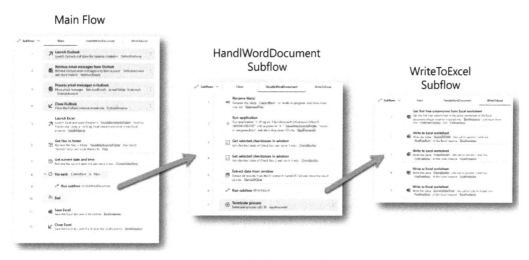

Figure 3.9 – Main flow and subflows

A subflow is stored within the flow itself and can only be used in the containing flow. If you look at the **Actions** pane, you can also see that there is a **Run flow** actions group with a **Run desktop flow** action in it. This additional option will be discussed in *Chapter 4*.

To create a subflow, click on the **Subflows** drop-down menu at top of the workspace, as shown in the following screenshot. In the drop-down menu, you can also see any already existing subflows:

Figure 3.9 – Creating a subflow

The new subflow appears as an additional tab next to the main flow and it starts with a blank workspace to drag actions into. Another option is to cut and paste one or more actions from the main flow over to the subflow. Within the main flow, you can now use the **Flow control | Run subflow** action to specify the subflow to run in the corresponding parameters dialog.

Now that we've learned some ways to edit flows, let's explore ways to use those skills to effectively create them.

Strategies for effective flow creation

We have already covered the flow designer window and the menu options in the last section. In this section, we want to look at how we can make use of these possibilities to incorporate them in the design of a UI flow.

Let's take our initial example from the very first chapter. The challenge was to automatically extract information from multiple Word documents sent to us via email. We want to save that information into an Excel spreadsheet for further reporting. In the end, this flow had about 22 actions, 13 flow variables, and 2 input/output variables. This flow already has a significant complexity, so let's explore how we can reduce this complexity and implement the flow.

Start simple

The first suggestion is to encourage you to start simply and work incrementally. Taking our given example, just start by dragging the **Launch Outlook**, **Retrieve email messages in Outlook**, and **Process email messages in Outlook** actions onto the workspace and play around with them. Once you have familiarized yourself with this topic, you can then move on to the next task. It is also possible to create separate flows to test things out and then take the knowledge to incorporate it into another flow.

Split the whole task

Building on the previous concept, it is often useful on larger tasks to break the overall task into several pieces, as already shown in *Figure 3.8*. You can do this by using subflows. Referring to our previous example, this could look like this:

- **Part 1 – message and attachment retrieval**: We need to retrieve all new messages from parents
- **Part 2 – processing of each attachment**: We want to look into each Word document and extract relevant information to store in the spreadsheet
- **Part 3 – cleanup and housekeeping**: Close all open programs

We can now work on separate tasks, which straight away reduces the complexity.

Work on a single item before looping

In our example, we want to process several Word documents sent to us by parents. We can use the XY action to access all documents in a list.

However, before we loop through the list of documents, we must first clarify for a single document what data we want to extract and how we want to store it. So it makes sense to go through the flow with a single document and design these steps before doing the whole thing in a loop for all documents. Later, it is possible to drag a loop action into the workspace and place the single processing there. To do this, you can simply drag and drop the required actions, as we learned previously in this chapter.

Deactivate to test steps

At some point, a longer flow will come up that will be run through very often to test it. The same steps will be run through again and again from start to finish. However, if only a specific part is to be tested or refined, it makes sense to omit those actions that are not in focus. These actions can then simply be deactivated for this purpose, and a breakpoint can be used to stop at the desired action.

In our example, we could run into an error when listing the documents in the **Get files in folder** action. For error detection, it is irrelevant whether Outlook was started before and all emails were fetched. Therefore, we can simply disable those very steps for this test. In addition, it makes sense in this situation to set a breakpoint in the **Get files in folder** action to pause the flow exactly where the error occurs or shortly before. We can see this condition in the following screenshot:

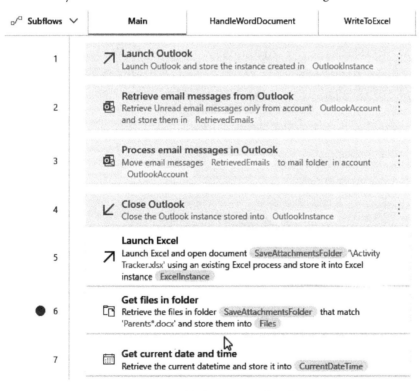

Figure 3.10 – Disabled actions and breakpoint

In this way, the techniques learned in this chapter can be used to develop and troubleshoot a UI flow quickly and efficiently.

Adapt to the available options

Sometimes you will have a great idea for a flow and how to make your daily tasks easier. But then you realize that, unfortunately, the flow can't be mapped and designed the way you imagine it because there aren't the right actions. In our example, we basically want to process every single email that reaches us from the parents, save the attachment and extract data from it. Although the **Retrieve email messages from Outlook** action offers us, in a variable, all emails that are new and match our criteria, so we can run through this list of emails in a subsequent loop. However, unfortunately, in this loop, we cannot access the attachment of the email to analyze it and extract the data. So, we have to think of another way to get to the assets and extract the data. Fortunately, however, the very action mentioned previously contains a parameter that allows us to save all the attachments in one folder. Accordingly, if we break the flow down into smaller components and set it up so that we first save all the attachments and then go into individual processing at the document level, then we can also use the actions right out of the box.

Summary

PAD offers many possibilities to create and structure UI flows. The designer window additionally offers great functions to manage even larger flows well. Working with actions, easy moving, and deactivating together with the breakpoints offers great possibilities even for beginners to create flows efficiently and to find errors. Based on the strategies introduced in this chapter, we will continue with our journey and expand our knowledge by creating and controlling the flow even further in the following chapter.

4
Basic Structure Elements and Flow Control

A flow is typically executed from top to bottom. Sometimes, however, we need to include branches or even repeat certain actions. In this chapter, we will look at how the sequence of actions in processing can be controlled, because controlling the flow logic is an essential part of its creation.

In this chapter, we will cover the following topics:

- Using conditionals
- Handling repetitive tasks with Loop statements
- Error handling

Technical requirements

You will need a workstation with PAD installed on your desktop, and you should know how the application works and how to create a simple flow.

Using conditionals

Conditionals are used to ensure that a certain state is checked in advance before an action or a list of actions is executed. They are one of the basic concepts in programming and can be used to build a flow without programming knowledge. We already used conditionals in one of the previous examples where we needed to check whether a folder existed before we created one. Conditionals are commonly referred to as `If` statements.

To use a conditional, we need to drag one of the actions in the **Conditionals** action group onto the workspace. We can locate this action group at the top of the list, as shown in the following screenshot:

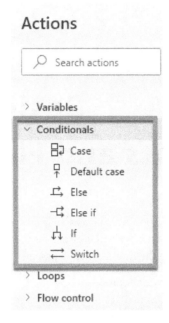

Figure 4.1 – Elements in the Conditionals action group

It should be noted that these actions, listed in *Figure 4.1*, must be used in a specific order and sequence. The following subsections explain how to use each action in context.

Starting with If and Else

In most cases, when we need to check a condition, the If statement action is the starting point. If you drag an If action onto the workspace, the actions parameter dialog appears and needs information, as we can see in *Figure 4.2*:

- **First operand**: Determines what we want to check
- **Operator**: Determines which operation to use for the check
- **Second operand**: Determines what to check against

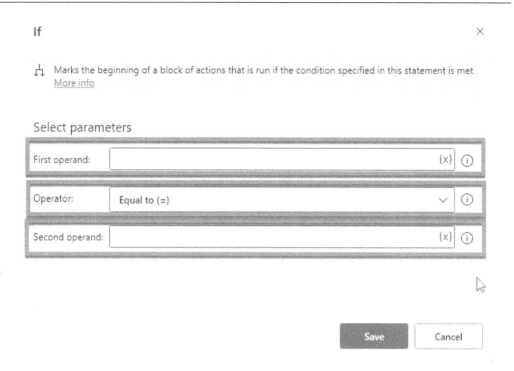

Figure 4.2 – If action parameter dialog

Essentially, what happens here is that we can compare two values (first and second operands) with each other (equal or not, greater than, etc.) and then decide what should happen if they do or don't match. If we now populate the first and second operands with some meaningful values (or some static text for demonstration purposes), the workspace will look like this:

Figure 4.3 – A generic If block

As we can see, an End action has automatically been placed below the If action to mark the end of this block.

> **Do not disable or delete the End block**
>
> The End closing block is just another action in that it can be moved around, disabled, or even deleted. If you do this, the flow designer immediately shows an error saying that the End statement is missing. If you accidentally delete the End statement, you can just go to the **Flow control** actions group and drag a new **End** action to the right place in the workspace. Be careful as well to put End at the end of your conditional block and not just after If.

We can now drag additional actions right inside this block. These actions will only be executed if the If action and the operator evaluation will be true. For example, we could simply drag a **Display message** action into the block to display a message box if the evaluation returns true. But what if the result is false? Well, in this case, nothing would happen unless we also drag an Else action into the workspace between the If and End statement. All actions that are now placed below the Else block will be executed when the If block evaluates to false. To test this out, we could also drag a **Display message** action into the workspace below the Else block and display a different message, as shown in the following screenshot:

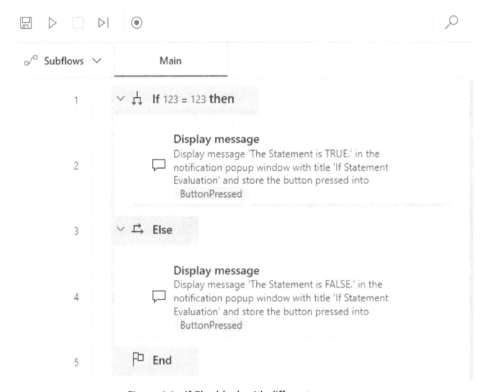

Figure 4.4 – If-Else block with different messages

Another important concept is that If statements can be nested; that is, within an If block, it is possible to place another If block. We will learn some examples of this in the course of the chapter.

As it does not really make sense to compare two static values, we would want to compare values that are stored in variables. We will dive deeper into the topic of variables in the next chapter, but for the purpose of this chapter, let's create some simple variables. To do this, use the *plus* button on the right side of the workspace in the **Input / output variables** area, and choose **Input**.

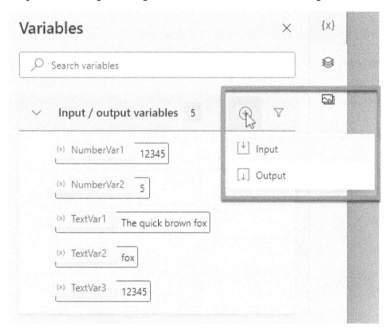

Figure 4.5 – Create a new input variable

In the following dialog, we have the option to give the variable a name, a data type, and a default value. Use this dialog now five times to create the following variables:

Variable name/External name	Data type	Default value
NumberVar1	Number	12345
NumberVar2	Number	5
TextVar1	Text	The quick brown fox
TextVar2	Text	fox
TextVar3	Text	12345

Table 4.1 – List of variables for the conditional test

The following screenshot shows the **Edit input variable** dialog to create a variable, in this case, for NumberVar1. Click the **Save** button to store the variable.

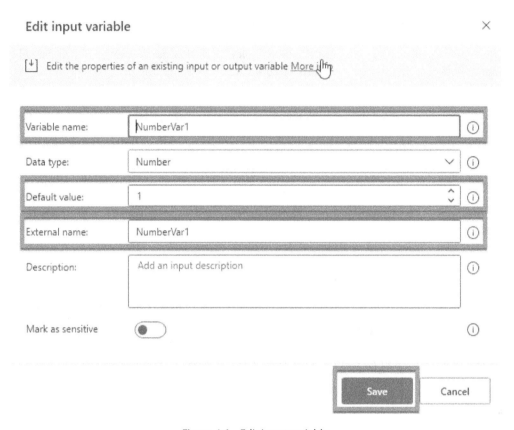

Figure 4.6 – Edit input variable

In the end, you should see all variables as in *Figure 4.5*. We can now replace the static values in the If statement from the start with these variables. To do this, double-click on the If block to bring up the parameters dialog again. Now, we can edit the first and second operands with some variables:

1. Delete the existing static values (in our example, 123).

2. Press the {x} button on the right side of the input box. This launches the subwindow to select a variable. You should see all the variables you entered in the previous step.

3. Select **NumberVar1** for the first operand.

4. Press the **Select** button.

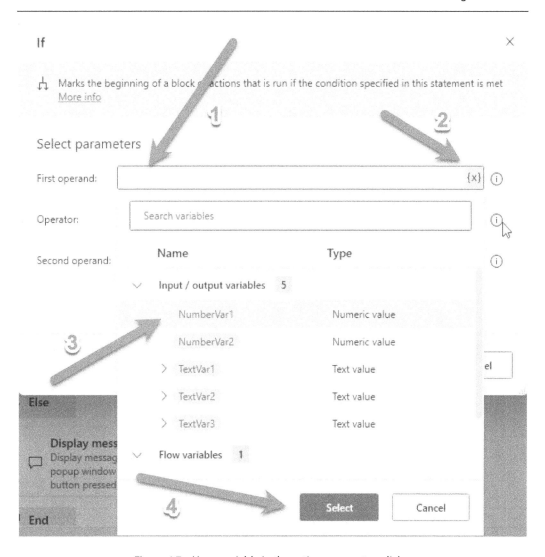

Figure 4.7 – Use a variable in the action parameters dialog

Repeat the steps for the second operand and choose **NumberVar2** as the variable. The final result is displayed in the following screenshot:

Figure 4.8 – Variables in the parameters action dialog

We can now very easily play around with different variables and content to understand better how the different operators behave. It is also possible to use different types of variables for operand 1 and operand 2. The **Operator** dropdown now determines how to compare the two values. The result of this comparison is true or false. Only if the result is true will the actions inside the If and corresponding End container be executed. Here, we have the following options:

- **Equal to** (=): Returns true if the two variables are identical.

- **Not equal to** (<>): Is the opposite of **Equal to** and returns true if the two variables are not identical.

- **Greater than** (>): Returns true if the first operand is a higher number or a longer text than the second operator.

- **Greater than or equal to** (>=): Just like *Greater than*, but also returns true if the two operands are identical, not only when the first operand is a higher number.

- **Less than** (<): Returns true if the first operand is a lower number or a shorter text than the second operator.

- **Less than or equal to** (<=): Same as *Less than*, but also returns `true` if the two operands are identical.

- **Contains**: Returns `true` if the second operand is contained in the first operand; for example, `fox` (second operand called `TextVar2` as a variable) is contained in `the quick brown fox` (first operand, `TextVar1`, as a variable). This also works for numbers, for example, if `NumberVar2`, with a value of `5`, is contained in `NumberVar1`, with a value of `12345`. It even works with mixed types of variables, say if `NumberVar2` is also contained in `TextVar3`.

- **Does not contain**: Is the opposite of **Contains**.

- **Is empty**: Returns `true` if the variable does not contain any value. This operator works with only one operand. To test this, just remove the text from one of the text variables and then using this variable with `is empty` condition. This does not work for variables of the number or Boolean type, because these types always get a default value

- **Isn't empty**: Is the opposite of **Is empty**.

- **Starts with**: Returns **true** if the value in the second operand is contained at the beginning of the first operand. If you change, for example, the value of `TextVar2` to `the` and use this operator with `TextVar1`, the result will be `true`.

- **Doesn't start with**: The opposite of **Starts with**.

- **Ends with**: Same as **Starts with**, but compares the end of the values.

- **Doesn't end with**: The opposite of **Ends with**.

The operators which can work with operands of type text also allow ignoring the upper and lower case of a value. These include **Contains / Does not contain**, **Is empty / Isn't empty**, as well as **Starts with / Doesn't start with / Ends with / Doesn't end with**.

The result of the comparison is largely dependent on the content of the variables. I want to encourage you to rebuild the simple flow in this example and try different options, changing the operator and the values of the variables. By using the **Display message** actions, you immediately see the result of the operations. PAD is very friendly here and also allows the use of variables of different types.

Use Variables of the Same Type

Although the Microsoft documentation on conditionals (`https://learn.microsoft.com/en-us/power-automate/desktop-flows/use-conditionals`) states differently and you won't get an error when comparing a variable of the text type with a variable of the number type. For this operation, both variables need to be converted to the same type. The result of this conversion is not always predictable and even if the result would make sense in some cases, it is recommended to use variables of the same type to avoid inconsistencies or unexpected behavior.

Else If action

Sometimes one conditional action is not enough to branch to the right path of actions that need to be executed. With the regular `Else` statement, the actions in that section are always executed as the only one alternative branch because the condition is not evaluated again.

With `Else If`, it is also possible to define another conditional in the `Else` part, which can also be totally independent of the first conditional. To do this, simply drag the `Else If` action between the beginning `If` block and the `Else` block, which must always be the last block in this construct. The following example should illustrate the use of conditionals. The idea is the following:

- It should be evaluated whether a hockey team has won a game, and if so, what the goal difference was.

- Depending on the values, different message boxes will be displayed.

- If the team did not win, it should be evaluated how many fouls there were. If this number is low enough, the team will be rewarded with a special fairness message.

I created four variables to store the values:

- `GoalDifference` (number)

- `ManOfTheMatch` (text)

- `NumberOfFouls` (number)

- `TeamGameStatus` (Boolean, which can be `true` = won or `false` = lost)

This is what the flow looks like:

Figure 4.9 – Hockey match evaluation flow

Please note the following:

- This is just an example implementation, where a team can get the winning message with special acknowledgment *or* the fairness message. This is just to demonstrate the use of the `Else if` statement and does not mean that a team cannot win and also be very fair, of course.

- Only one message will be displayed when this flow is running. The flow branches exclusively to the section for which the evaluation of the condition is true. If none of them fits, it branches to the `Else` branch. You can use many `else if` inside the loop,

- You can easily change the variables and their values to test out the different branches.

Conditionals can be very helpful when designing flows. The simplicity of creating a UI flow with PAD allows you to jump right in and design. In a more complex scenario, however, it is also very useful to step back and draw the logic in a flow chart. Multiple nested conditionals with `Else if` can especially be very hard to maintain later, so it is always a good idea to simplify the logic as much as possible.

Special If actions

The conditional concept described in the previous section is able to deal with hardcoded values and variables. But what if we need to do some more sophisticated things, such as checking whether a file exists?

PAD offers special `If` actions for these purposes and we will cover them in the corresponding sections. To get a first impression, here is a list of special `If` actions with a brief description of their functionality:

Name of the action	Located in action group	Functionality
`If file exists`	Files	Checks whether a file exists or does not exist
`If Folder exists`	Folder	Check whether a folder exists or does not exist
`If service`	Windows services	Checks whether a service is stopped, installed or not installed, running, or paused
`If process`	System	Checks whether a process is running or is not running
`If window`	UI automation	Checks whether a window is open or not and whether it is the foreground window
`If window contains`	UI automation	Checks whether a specific piece of text or UI element exists in a window
`If image`	UI automation	Checks whether a selected image is found on the screen or not
`If web page contains`	Browser automation	Checks whether a specific piece of text or element exists on a web page
`If text on screen (OCR)`	OCR	Checks whether a string / sequence of characters appears on the screen or not

Table 4.2 – List of special If actions

As we can see in the preceding table, PAD offers a lot of functionality to leverage a conditional sequence of actions in our flows. It is even possible to have multiple branches for conditionals, which is explored in the following section.

Conditionals with multiple branches

Sometimes there is also a need to classify a value into one of several categories rather than perform a single test. A school grade could be determined, for example, on the basis of a certain percentage achieved. There are several branches, each of which could list its own sequence of actions. This is where we would use the `Switch-Case` construct.

Figure 4.10 – Evaluating student grades with Switch-Case actions

In the preceding example, we can see how student grades could be determined by using a `Switch-Case` construct. These actions are also located in the **Conditionals** action group. In the preceding example, we want to create a flow that displays an individual message box depending on the percentage of the grade according to the following rules:

- `Grade A > 90 - 100%`
- `Grade B > 80 - 89%`
- `Grade C > 70 - 79%`
- `Grade D > 60 - 69%`
- `Grade F > 0 - 59%`

To create the preceding flow, follow these instructions:

1. Create an input flow variable called `StudentGradePercentage` of the `Number` type and give it some default value, in my case, `85`.

2. Locate the **Switch** statement in the **Conditionals** action group and drag it onto the workspace. Notice that there is also a corresponding `End` statement created.

3. In the actions parameter dialog for the action, use the variable from *step 1* as a value to check.

4. From the **Conditionals** action group, drag a **Case** action between the `Switch` and the `End` actions.

5. Use `Greater than or equal to (>=)` as an operator and the value `90` as a value to compare.

6. Find or locate the `Display message box` action by using the search bar or the **Messages** actions group and drag the action right under the `Case` action.

7. In the dialog for this action, enter a title and a message according to the grade.

8. Repeat *steps 4* to *7* with different values and messages for grades A to D.

9. For grade F, we use the `Default case` action, which is the last action in the whole statement. We don't have to specify any value for this. If none of the previous conditions are `true`, this action will be executed. Don't forget to also place **Display message** here.

Please note that within a `Switch` statement, only one branch is executed at a time, even if a value may apply to multiple conditions, and it will be the first one that will match the condition. The sequence plays a decisive role. For example, if I were to list the `Case >= 80` instruction first, the value `95` would fall in here, the corresponding action would be executed, and the construct would be left, although actually, another branch (`Case >= 90`) would have been the correct one. We have now learned how we can control the sequence of actions, depending on certain conditions, be it with a single branch or with several branches. The next section now takes care of what we need to do to be able to repeat certain actions depending on the parameters.

Handling repetitive tasks with Loop statements

Another valuable concept in programming languages is to iterate through a list or run through a specific set of statements and actions a defined number of times. This can also be achieved in PAD by using loop actions. These are located in the **Loops** action group.

The Loop action

Let's start with the `Loop` action and create a dice simulator for the *Ludo* board game. In this board game, players have to circumnavigate the board using pieces of a color chosen before the start of the game. The dice is rolled three times. If the result is a 6, the player can move a piece. If not, they need to try again. Here is the flow to simulate this situation:

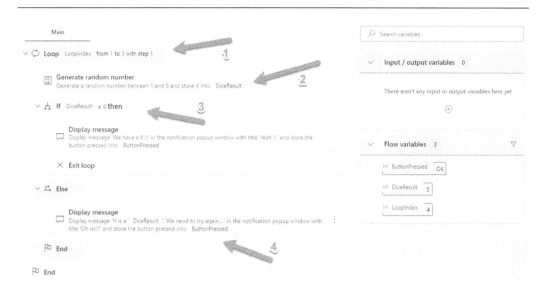

Figure 4.11 – Roll the dice flow with a loop statement

The following is a detailed explanation of the steps you see in the previous figure:

1. Open the **Loops** action group and locate the Loop action. Drag it onto the workspace and set the following values:

    Start from: 1

    End to: 3

    Increment by: 1

    This will cause the loop to be executed three times.

2. We now need to create a random number as a result of rolling the dice. Use the search bar above the **Actions group** section and search for the random keyword. You will find an action called **Generate random number**. Drag it right under the Loop action. In the action parameter dialog, enter 1 for the minimum value and 6 for the maximum value. Try also to rename the default variable name from RandomNumber to DiceResult by expanding the Variables produced area and double-clicking the name. Make sure that the new variable name is surrounded by the percentage sign, like this: %DiceResult%.

3. Now, we need to check whether the result is equal to 6, which means we would continue with the game and stop rolling the dice again. We will now use the conditional from the last section in this chapter to check the value for DiceResult. If the result is equal to 6, we will display a message box confirming this.

> **Important Note**
> After the message box, we use the `Exit loop` action to stop rolling the dice again. This action is also located in the **Loops** action group and doesn't need any additional configuration.

4. In the `Else` branch of the `If` action, we just want to display the result of the dice if it is not equal to 6.

Please note that in the preceding example, we used the `Exit loop` action to leave the looping procedure. There is also a `Next Loop` action, which can be used inside a `Loop` action. Depending on where the action is placed inside the `Loop` action, all following actions will be skipped, and the next iteration is forced.

Loop condition

Another way to loop through a list of actions is by using the `Loop condition` action. In a programming language, this is referred to as a `while` loop. While a specific condition is `true`, a set of actions will be executed repeatably.

In the next example, we will see how this works. The following flow implements a guessing game. At the beginning, a random number is determined, and the user has to guess which number it is. If the user enters the correct number, the game is won and the loop ends; otherwise, the number of attempts is increased. Here is the implementation for the flow:

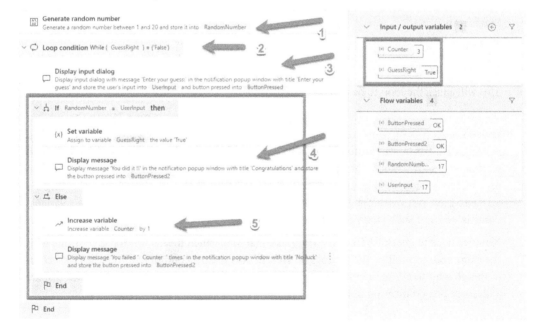

Figure 4.12 – The guessing game implementation

Here are the steps explained in detail:

1. The flow starts with the definition of two variables, Counter (a number, initialized with 0) and GuessRight (a Boolean value initialized with False, storing the status of the user's guess), and a Generate random number action. To make it a bit easier for the user, a number between 1 and 20 will be generated and stored in the RandomNumber variable.

2. Next, the Loop condition action will be used. The first operand in the actions parameter dialog will be the GuessRight variable, which we want to check against the fixed value of False. This means as long as the value of GuessRight is false (the user did not enter the right number), the loop will continue.

3. Next, we want to display an input screen that allows the user to enter a number. We use the Display input dialog action for this, which is also located in the Message boxes actions group. We only have to enter a title and a message here. The input value will be stored in a variable called UserInput.

4. After the user has entered the guess, we need to check whether the random number is equal to the user input. We use an If conditional for that. If the two values are the same, we use the Set variable action to modify the GuessRight variable to true and display the success message. By modifying the variable, we also cause the loop to end.

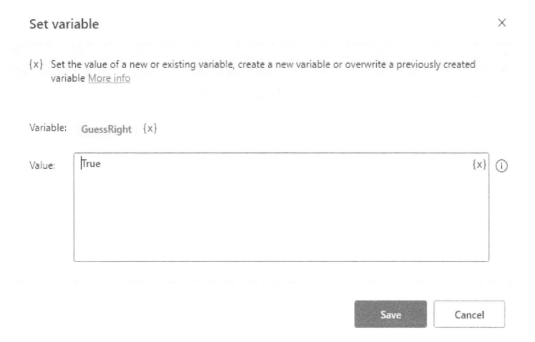

Figure 4.13 – Set variable action

5. In the Else branch, we just increase the counter of the attempts of the user, as shown in *Figure 4.12*, and display a message box.

Spoiler alert: You can see the number that you need to guess in the **Flow variables** section on the right side of the window.

For each

Also, a very common scenario is to loop through given elements, for example, the files in a folder, as we have already seen in one of our previous examples. The For each action is used for this purpose and needs a list of items, a data table, or a data row as input so that a set of actions will be executed for each of the elements.

There are built-in actions that already produce a list of items, for example, Get files in folder or Get subfolders in folder, but it is also possible to create your own lists of items as variables and work with these.

We will discuss the topic of variables and lists in detail in the next chapter and see examples of how to use the For each action.

Additional action for flow control

In the previous sections, we learned about conditions and loop constructs, the basic building blocks for controlling flows. In this section, we will look at additional ways to influence the flow of actions in a flow. These actions are contained in the corresponding action group, Flow control.

Figure 4.14 – Flow control actions

Let's see what these actions can bring to the table. We will also see these actions being used in an example:

- **Comment**: This action is useful to bring comments into the flow and document the actions.

- **End**: Does not work on its own, but if we accidentally delete an `End` action that was part of an `If` action or a loop, we could just use this action to complete the flow.

- **Exit subflow** and **Run subflow**: We already used **Run subflow** in one of our earlier examples. **Exit subflow** can only be used in a subflow to stop the subflow and return to the point it was called from.

- **Get last error**: Retrieves the last error that occurred in the flow and stores it in a variable for further processing.

- **Go to** and **Label**: **Label** sets a marker in a flow, which is a specific position in the list of actions. With **Go to**, a label can be used to transfer the flow of execution to it. For example, it is possible to drag the **Label** action anywhere in the flow and give it the name `Position1`. With **Go to**, we now could say `jump to the label called` **`Position1`** `and execute from there`.

- **On block error**: Allows us to encapsulate a set of actions in a block and capture errors in execution eventually. We will discuss this in the next section.

- **Stop flow**: Stops the flow immediately when this action executes. It is possible to configure whether the flow has ended successfully or with an error message. If the latter, it is also possible to define the error message.

- **Wait**: This action pauses the flow for a specific amount of seconds. This can be especially helpful if the flow needs to wait for a program to start or for a screen to refresh.

Let's use our guessing game from previously to make an alternative implementation without a loop statement but leveraging our additional flow control elements. The flow works exactly the same way and looks like this:

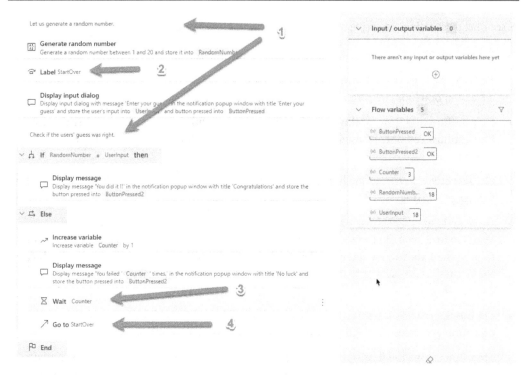

Figure 4.15 – The second guessing game implementation

A detailed explanation of the process is as follows:

1. Here, we have inserted comments to document the flow a bit more.

2. A `Label` action called `StartOver` was placed here before the input dialog. This is the position of the flow from where we want the execution to start again if the answer was wrong.

3. The `Wait` action pauses the flow for a certain number of seconds. To make this a bit more interesting, we can use the counter for failed attempts as the number here. That means the more attempts a user needs, the longer will it take to get the new message box for the next guess.

4. The `Go to` action jumps back to the `StartOver` label and executes from there.

As you can see, it is possible to create the same logic without a `loop` statement. The execution will only jump back to the top if the user does not guess right.

While there may be different styles of how logic is implemented, I would like to offer a word of caution at this point. The frequent and extensive use of `Go to` and `Label` statements can also quickly lead to confusion and also does not contribute to the easy readability of logic. Therefore, these constructs should always be used with care.

Now we have learned the two essential mechanisms for controlling the flow of actions. In the last section of this chapter, we want to turn our attention to error handling. This is both important and interesting because we deal with external input in our flows and can never be sure that the data we receive is in the right format or has the right content. So, how can we ensure that a flow will continue to run even if an error occurs? The following section shows us how this can be realized.

Error handling

Errors happen all the time. A file was not found, a folder does not exist, a UI element is missing – all these events could occur, and they will result in an error. Without further intervention, the flow would stop at the point where the error occurred. All further actions will not be executed. But this is often not what should happen. As a flow designer, we would want to check that error, maybe retry and log specific information, and execute the flow with some other actions until it finishes. This is what error handling is for.

On almost all action parameter dialogs, there is an **On error** button in the lower-left corner. This, of course, only makes sense if the action does something that could potentially go wrong. A Comment action or a Label does not have such an option.

We will see a lot more complex scenarios, but for this first encounter, we want to look at something simple: deleting a file that is not there. Let's create a new flow and define a simple variable, FileToDelete, of the text type and a default value of C:\deletethis.txt.

Next, locate the Delete file(s) action in the **File** action group and drag it in onto the workspace. The action parameter dialog appears and we can put in the variable name as the file to delete. But before we press **OK** now, we would want to use the On error functionality to handle the error. If we press the corresponding button in the dialog, it should look like this (I expanded the dialog a bit) on the left side:

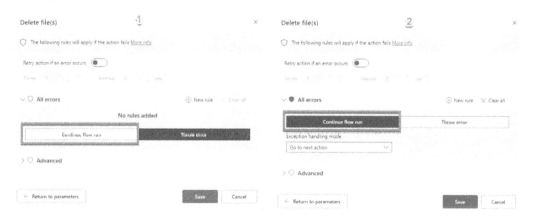

Figure 4.16 – On error dialog

There are two main options we can choose here:

- **Throw error** (right side): This is the default setting and will stop the flow from running.
- **Continue flow run** (left side): If we click on this option, the dialog changes to the one on the left side in the preceding screenshot. Now we can decide what should happen if an error occurs.

When we choose to continue with the flow run, we can determine with the drop-down box below this option what should happen. Here, we have the following options:

- **Go to next action**: The error will be logged, and the next action will be executed
- **Repeat action**: The failing action will be repeated
- **Go to label**: A label can be chosen to jump to

At this point, the flow does not stop with an error but continues to run according to the settings. But the **On error** dialog has even more options. Right at the top, there is an option to repeat the action a certain number of times and with a defined seconds interval. This can be used in addition to the **Wait** action. Imagine if some other external processes are creating this file and need longer than expected. Instead of failing, we could use this parameter to give it another try.

When an error occurs, it might also be helpful to log this somewhere or set specific variables with some information. This can be achieved by defining a rule by pressing the **New rule** button in the dialog. We can set a specific variable to a specific value, or we could even run a subflow to do additional error handling or capturing.

The **On error** dialog is basically divided into two parts: one is for handling all errors and the other one is for handling action-specific errors.

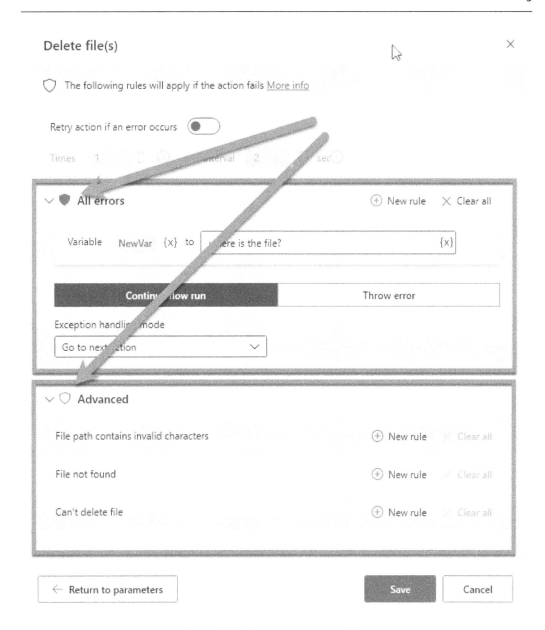

Figure 4.17 – Advanced error handling

Only one section can be active, which can be identified by the blue shield icon shown in *Figure 4.17*. The **Advanced** section contains all errors that can occur in conjunction with the file that should be deleted. If we were to take another action, these options would change accordingly. But this allows checking for specific errors for an action.

In my example, I choose to use the **All errors** section and set a new variable called **NewVar** with some additional information. Pressing **Save** here and designing the rest of the flow looks like this:

Figure 4.18 – A flow with an error

Notice that the first action, **Delete file(s)**, now has a new icon to the left of the line number indicating that error handling is active. We can also use the **Get last error** action to retrieve the error itself and store it in another variable, which we display in a message box afterward.

Now we know how to provide error handling for an action. Now, let's look at what we can do if we don't know exactly where in a list of actions an error occurs and how we can catch it.

On block error

We saw in an earlier example that it is easy to define error handling for an action and let the flow run to its end, even if an error occurs. But what if we have multiple actions that could fail? Or even a specific set of actions with some logic that might run into an error? The On block error action could help here. This action allows placing one or more actions inside this block and defining what should happen in case of an error. This is very similar to the error handling we looked at previously, except that there is no advanced section for action-specific errors. In programming, this is referred to as a **try-catch block**.

The same example as previously but now with the On block error action would look like this:

Figure 4.19 – Error handling with Onblockerror

The definition of the action parameter dialog of On block error is the same as for the individual action. The flow will run to the end and the message box will display the corresponding error message.

Summary

Being able to control the flow execution is essential for a flow designer. In this chapter, we learned how to use conditionals and loops together with additional ways to control the flow. Lastly, we looked at how to deal with errors in a flow and bring the flow to an orderly end.

Throughout the rest of the book, we'll keep applying what we've learned in this chapter, and that's equally true in the next chapter, where we'll learn about another important concept: variables.

5

Variables, UI Elements, and Images

Normally, a flow works with input from outside, be it values that are passed or applications that need to be controlled remotely. In this chapter, we will look at these input elements. These are located on the right-hand side of the designer window, which is also where we can find the areas for variables, UI elements, and images (see *Figure 5.1*). We will cover the following topics in detail:

- What exactly are variables and what kind of variables are available?
- What are UI elements and how will these be used in a UI flow?
- What role do images play in the context of PAD and how can they be used?

What these three topics have in common is that they can all be used as inputs for automation.

Variables and how to use them

A **variable** is an object (or design element) in which a value can be stored. You could also think about it as a reusable component that can store any value. There is no meaningful flow without at least one variable. Variables are available and can be created on the right-hand side of the designer window, as shown in the following figure:

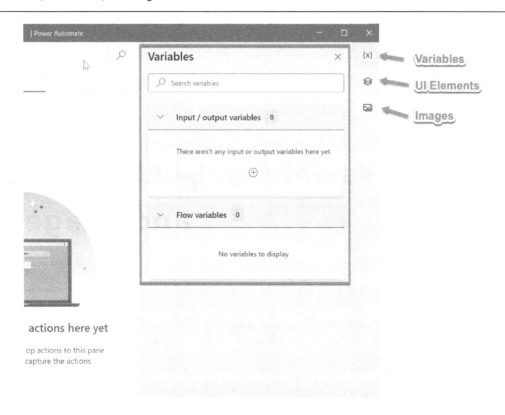

Figure 5.1 – Variables in the designer workspace

On the far right-hand side, there are also symbols for UI elements and images, which we will also cover in this chapter. With this pane, we can search for variables, create new input/output variables (see the next section), and also inspect variable values. The following figure shows the same area, which is called the **Variable value viewer** area, with some populated values and additional features:

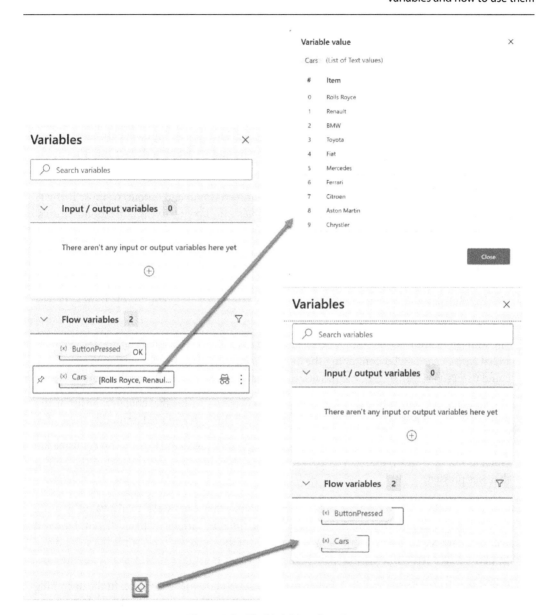

Figure 5.2 – The Variable value viewer

This pane not only shows the current values of the variables directly next to the name when the flow runs, but we can also inspect these values by double-clicking on their names. In this case, **Cars** is a list with the values shown in the popup dialog. The flow also maintains the values when it finishes until the next run starts. If we want to clear the values of all variables, we could use the little rubber icon, at the bottom.

Variables always have a specific type, such as **text** or **number**. This determines what kind of data can be stored in that variable. There are multiple types available, and we will dig deeper into this in the subsequent section.

We can distinguish between variables that we create on our own (input and output) or those that will be created by actions (flow variables). Each of these variables will show up in the corresponding area. Let's take a look at each of them.

Flow variables

This group of variables is produced by a flow action so that it can be used later in the flow. Let's take a look:

- The **Run application** action (the System group) starts a given application and creates a variable called AppProcessId containing the Windows process identifier for that application. We could use Terminate process with this variable to shut down the application.

- The **Get files in folder** action (the Folder group) creates a variable that contains the list of files that reside in that folder. This list can be used to iterate through and read or modify these files.

- The **Display message** action (the Message boxes group) shows a message box for confirmation (such as **Ok** or **Cancel**) and therefore creates a variable that contains the text of the button that was pressed.

Flow variables cannot be set directly in the designer. However, they can be assigned a value by using the Set variable action. Depending on the action, a different variable with a different type will be created. In the preceding list, the first variable would be of the number type, while the second one would be of the list type. This type cannot be changed, so we must take these variables as they are.

Let's create a new flow and just drag the **Display message** action into the workspace. In the upcoming dialog, take a look at the lower part in the **Variables produced** section and expand this with the down arrow symbol. It should look like this:

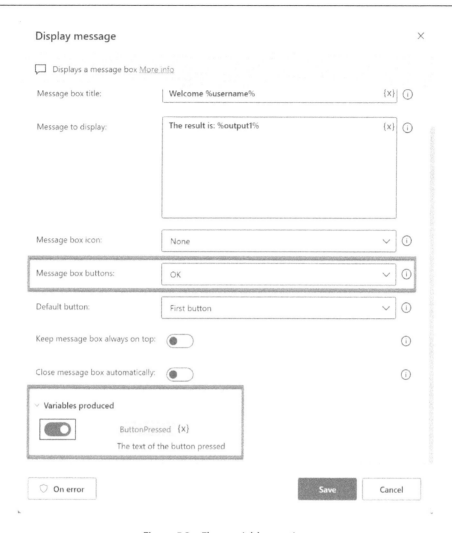

Figure 5.3 – Flow variable creation

Since this action displays a message and provides buttons to dismiss the dialog, a variable called **ButtonPressed** is created. Please notice that I also changed the value for **Message box buttons** to **OK – Cancel**, which will not just give us one **OK** button in the dialog, but two. The value for the ButtonPressed variable will contain the name of the button that the user will press: **OK** or **Cancel**. With this, a flow designer could evaluate that value and perform actions, depending on the button that was pressed. The examples provided by Microsoft make heavy use of this functionality. As a reminder, you can access these examples in the main PAD window and click on **Examples** to the right of **My flows**:

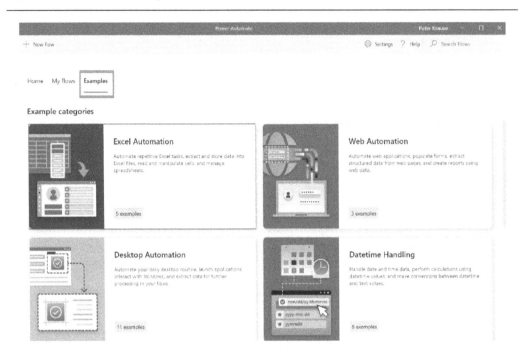

Figure 5.4 – Examples for PAD provided by Microsoft

Also, notice that we could disable the production of a variable with the corresponding toggle switch. This applies to all actions of PAD that could expose variables. If we are not interested in the content of the variable and won't need this in further processing, this would also make sense and keep the list of variables clearer (see *Figure 5.3*).

Regarding clarity, it is also possible to reuse a flow variable. Let's assume that we need another message box later in the flow. Dragging another action into the designer would produce another variable – in this case, `ButtonPressed2`. If we don't want to preserve the result from the first message box, we could then rename the variable `ButtonPressed`. To do this, we can double-click the name of the variable in the dialog and change it. In this way, we can overwrite the name of the existing variable instead of creating a new one. However, we can also use another method, which we can find in the kebab menu, along with other actions, as shown in the following figure:

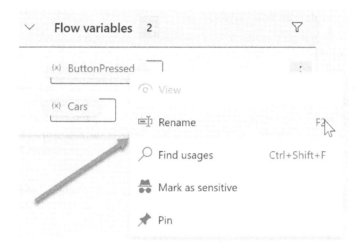

Figure 5.5 – Options for a flow variable

The following actions can be selected; each refers to the selected variable:

- **View**: Displays a dialog with the current value of this variable.

- **Rename**: Gives this variable another name.

- **Find usages**: Searches the flow for this variable and displays the result in a separate window in the lower area of the window. It's a very convenient function to find all places where a variable is used.

- **Mark as sensitive**: See below for further explanation.

- **Pin**: Pins this variable to the top of the window. All additional variables will be shown below the pinned variables.

If a flow variable is renamed to an existing name, we need to confirm that these two variables should be merged. If we take our example with two message boxes from earlier and rename **ButtonPressed2** to **ButtonPressed**, this is what it looks like:

Figure 5.6 – Merging flow variables

> **Merging is only possible with flow variables**
>
> The merge functionality is only available for flow variables and not for input/output variables. To change the name of an input/output variable, you can just edit it and change the name, but the name must be unique between all other input/output variables. Otherwise, an error message will be shown directly in the variable editor. However, a flow variable could have the same name as an input/output variable. Flow variables cannot be deleted explicitly. These get removed when the corresponding action (dependency will be set automatically) is removed from the flow. If, however, an action reuses a variable, as in our example, the flow variable will be kept until the last action is removed that uses this variable (this is because of security reasons.).

Flow variables also disappear if the corresponding action is disabled. Any action that uses this variable later in flow logic will fail.

Input and output variables

We made use of input variables in our previous examples. Input and output variables enable us to bring in data from the outside into the flow (input variable) and also expose data to the outside (output variable). The latter especially makes sense if the flow is part of a larger construct and needs to return some values to another cloud flow, for example.

We used these kinds of variables in the previous examples to define some input data by setting the values manually (we can set them as default values when we define the variable), which could be very helpful in some situations and is not possible with flow variables. When we need the same data in different locations in a flow, a variable should be used for this data.

As an example, let's assume we want to create a flow that does some complex folder operations and file movement within a specific folder. This specific folder will be used in multiple actions in the flow. If we want to change this base folder for some reason, we would need to do this in many locations. Instead, a variable could be introduced for storing the base file path from where all operations will run. Changing this variable will result in it being updated in all locations where it is used.

It is very easy to create a variable. In *Figure 5.1*, there is a little + sign in the middle of the field for input/output variables. Clicking this gives us the option to choose between an input or output variable. The following dialog to create the variable is the same for both types:

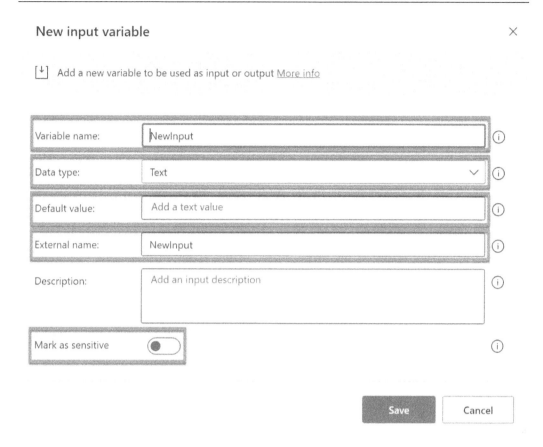

Figure 5.7 – Creating a new variable

We need to determine the following values:

- **Variable name**: The unique name of the variable without spaces.

- **Data type**: Determines the type of information that the variable will contain.

- **Default value**: The value that the variable will have when the flow starts. This is only available for input variables.

- **External name**: Will be used when the flow is called from outside of the designer window.

- **Description**: Allows us to enter additional text for a description.

- **Mark as sensitive**: These variables mask their values during debugging and runtime.

> **Important to know**
>
> Variables can be marked as sensitive and unmarked at any time. If marked as sensitive, the values of these variables are not encrypted or unreadable. The values can also be manipulated or used in some calculations or expressions. It is just a masking option for debugging or within the runtime.

So long as we are in the flow designer, we can just set the value of an input variable either while defining it or after by double-clicking the variable in the designer. But as soon as a flow gets started from the flow console, the **Flow inputs** dialog will appear and prompt us to enter values for all input variables.

Let's create a simple example to demonstrate this. Start with a blank flow, give it a meaningful name, and create the following variables:

Type	Variable Name	Data Type	Default Value	External Name
Input	op1	Number	1	Operand 1
Input	op2	Number	1	Operand 2
Input	Username	Text		Your name
Output	output1	Number		Return value

Table 5.1 – Input/output variables for a flow test

Now, create a flow that looks like this:

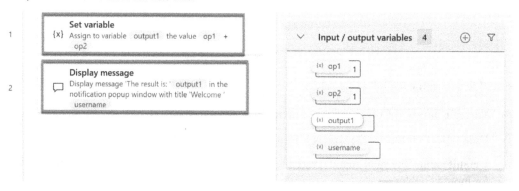

Figure 5.8 – A simple calculation flow

This is the list of actions for this flow, which has the following functions:

- Use the **Set variable** action to assign the result of the addition of op1 and op2 to the output1 variable. Notice the % notation, which we cover later in this chapter:

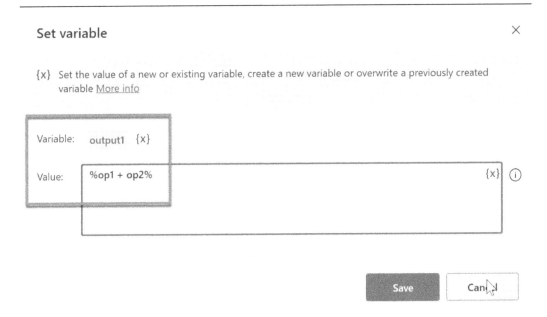

Figure 5.9 – Using Set variable with the % notation

- Create a **Display message** action, use the `username` variable in the title, and display the result of the calculation in the message text:

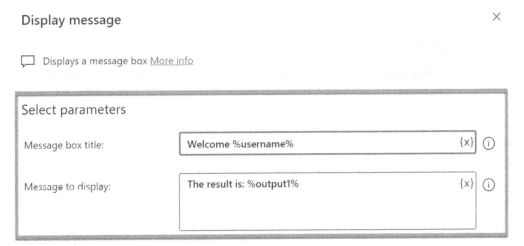

Figure 5.10 – Displaying the calculation result

Please notice that you can either write down the variables or use the **{x}** button to access the variables viewer. Now, save and close the flow designer and start the flow from the console window by pressing the **Play** button. You should see the following dialog:

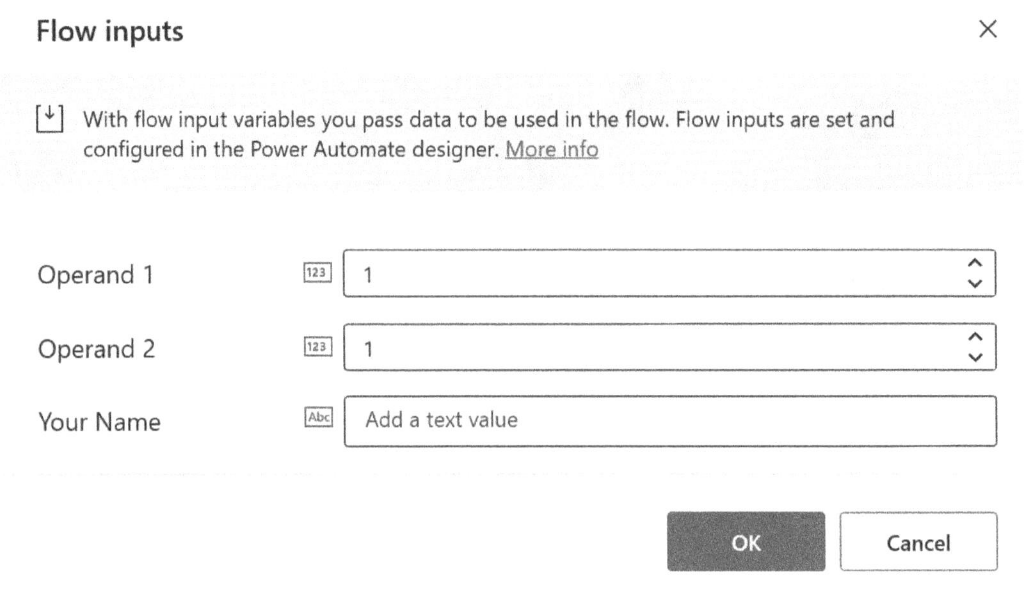

Flow inputs   ✕

[↓] With flow input variables you pass data to be used in the flow. Flow inputs are set and
     configured in the Power Automate designer. More info

Operand 1	[123]	1
Operand 2	[123]	1
Your Name	[Abc]	Add a text value

OK   Cancel

Figure 5.11 – The Flow inputs dialog

You will see the external variable names for the different inputs. If you enter some values for the operands and a name, the flow will continue to execute and display the message box and the output value.

Variable types

Input/output variables can have a simple type (text, number, or boolean) or a complex type (custom object, list, or data table). While a simple type can contain only one value (text = "some text value", number = 5, or boolean = True or False), complex types can contain multiple entries or even records. This makes sense, especially if desktop flows are part of a larger scenario, in which not only simple values but lists and tables are to be transferred and processed. Let's explore which complex variable types are available.

To work with variables, we have a dedicated actions group called **Variables** at the very top of the list of action groups. Here, we will find a lot of actions that deal with complex data types, but also some for the simple types, such as the number data type:

- **Truncate number**: Allows you to extract the integral or fractional part of a numeric value.

- **Generate random number**: Useful for creating a random number within given boundaries. We used this in the previous chapter in the dice simulation.

- **Increase variable** and **decrease variable**: Does what it sounds like.

The remaining actions are related to complex data types, which we'll look at next.

List data type

If we need to store multiple values separately, we could use a list. In programming terms, this would be a one-dimensional array. A list can be created by using the corresponding action, **Create new list**, or defining an input/output variable with that type. Other actions also create lists, such as **Retrieve email messages from Outlook** or **Get files in folder**. When you have a list, it is very convenient to iterate through all the items by using a `for each` loop. We saw this in previous chapters. Nevertheless, it is also possible to access individual members of a list through an index that starts with 0 as the first element. The syntax is as follows:

- `%Listname[StartIndex:NumberOfEntries]%`

  Example: Imagine a list of cars with the following entries: `Mercedes, Aston Martin, BMW, Chrysler, Fiat, Toyota`, and so on. The `%Cars[0:3]%` expression would select `Mercedes, Aston Martin`, and `BMW`. Remember: we start to count at 0.

We have different additional actions that can be used with a list. To manipulate the content of the list, we could use the following:

- `Clear list`
- `Add item to list`
- `Remove item from list`
- `Sort list, Shuffle list`
- `Reverse list`
- `Remove duplicate items from list`

All of these are quite self-explanatory and do what their names suggest. They do not expose any additional variables.

Some actions need two lists to operate. These are `Merge lists, Find common list items` and `Subtract lists`. However, these actions create a new list, which is the result of the corresponding operation.

Data table

In addition to the one-dimensional list data type, we could also use a two-dimensional or tabular data table to store values. You can think of a data table as a simple Excel spreadsheet. There is a specific action that reads data from an Excel spreadsheet and returns a data table. These actions create and manipulate data rows that are also located in the **Variables** action group but in the `Data table` folder.

To create a data table variable, we could use the **Create new data table** option. The actions parameter dialog contains an **Edit** button for defining the table:

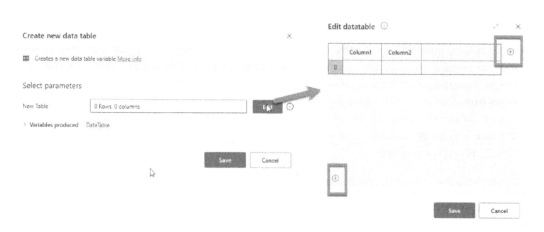

Figure 5.12 – Creating a data table

We could double-click on the existing columns to rename them or press the + button to add new columns and rows. **Save** will store the values and return us to the actions parameter dialog, indicating the number of rows and columns that the new data table contains. To demonstrate the different actions for a data table, I have created a simple table with six rows and two columns and called it CarsTable, as shown in the following screenshot:

Edit datatable ⓘ

	Manufactor	Country	
0	BMW	Germany	
1	Aston Martin	United Kingdom	
2	Chrysler	USA	
3	Fiat	Italy	
4	Citroen	France	
5	Toyota	Japan	

Save   Cancel

Figure 5.13 – The Cars table

Now, if we want to add a new row to that table, we could use the **Insert row into data table** action. In the corresponding action parameter dialog, we need to choose the data table and its **Into location**. Here, we can select **End of the data table** or **before a row index**, which needs to be specified when chosen. The new value we add must comply with a specific syntax. To add a new car record with Manufacturer = Chevrolet and Country = USA, we would need to enter `%['Chevrolet', 'USA']%`.

We will talk about the `%` notation in the next section, but essentially, we need to pass a list or a data row in the `New value(s)` parameter that is encapsulated by the percentage.

Update data table item works a bit differently because it is just updating one specific cell or item. We need to pass the column and row number as parameter values and the new value to set.

To delete a row, use **Delete row from data table** and pass the row index as a parameter.

The **Find or replace in data table** action is the most complex one because it has a lot of functionality. As its name suggests, it can just find or find and replace. It is possible to use regular expressions to find some text and specify whether to use a case-sensitive search. The result of this action is another data table that includes the columns and rows for the matches as separate records.

Let's take a look at the following example and see how this works.

Based on the Cars table, we want to insert a new row that contains a new car. Now, we want to find the text USA with `United States` in every place. We could create a flow that looks like this:

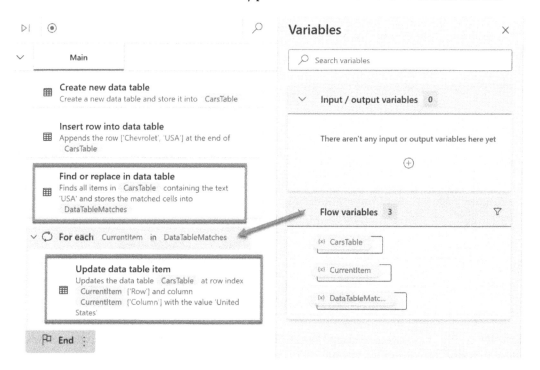

Figure 5.14 – Working with data tables

The **Find or replace in data table** action has been used here to find all occurrences of the term USA. I have configured it like this:

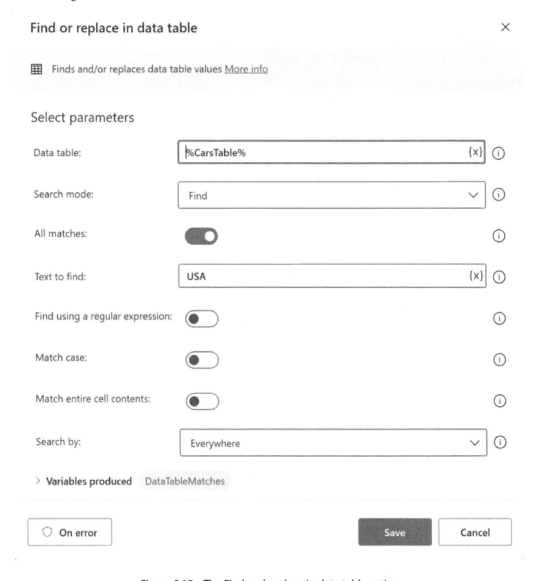

Figure 5.15 – The Find and replace in data table action

This action creates a new data table named **DataTableMatches**, which in our case contains two records, namely those for American cars. Double-clicking on this variable after the flow has run shows the content of this table:

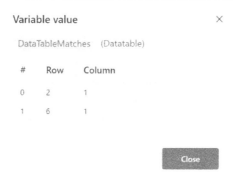

Figure 5.16 – Result table of Find or replace

This result table is always structured in the same way. It contains two columns named **Row** and **Column** containing the coordinates for the matches. In the next step, we iterate through this table and replace the text accordingly. The **Update data table item** action must be configured like this:

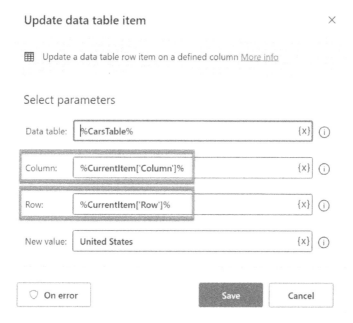

Figure 5.17 – The Update data table item action

As you can see, we can now use the column names of the result table and use these in the `CurrentItem` variable to address the right cell for the update.

You could use breakpoints to stop the execution and inspect the data tables. Comparing the `CarsTable` before and after the update would look like this:

Figure 5.18 – The data table before and after the update

Find and replace with "Replace" mode

Please note that we could have used the **Find and replace** action with "Replace" mode instead of just Find. By doing so, we would be able to determine a piece of text that should be used for replacement in the parameter actions dialog. That would be a much more convenient way to replace text in a data table and we would not need to iterate through the data table that was produced by this action. But sometimes, we only need to find a given piece of text without actually replacing it with some other text directly or doing even more complex things. The technique shown in the preceding example could then be used to accomplish these tasks.

This little example illustrates how data tables work. During this book, we will also use other actions that produce data tables. They are as follows:

- **Read from Excel worksheet** (the **Excel** action group)

- **Execute SQL statement** (the **Database** action group)

- **Extract data from web page** (the **Browser automation** action group in the Web data extraction subfolder)

These actions will be covered in the corresponding chapters.

Custom object

So far, we have looked at data types that have exhibited similar structures, such as a data table in which the associated records always had the same structure. However, we can also work with a data type called **object**, which is described by a set of properties and related values. Let's take the cars example from earlier in this chapter. For example, a car has the following properties: brand, color, year of manufacture, type of vehicle, and more. For some time now, the so-called **Javascript Object Notation (JSON)** format has prevailed here for the notation of such an object. This could look as follows:

```
{
    'Brand': 'BMW',
    'Color': 'blue',
    'year of manufacture': '2020',
    'type of vehicle': 'SUV'
}
```

It is possible to create an input/output variable for a custom object. All you have to do is select **Custom object** as the data type. Then, you can edit the data for the object in a separate dialog:

> **Tip**
> If you want to have a readable JSON file that you can easily edit, you can use the free JSON Formatter at `https://jsonformatter.org/`.

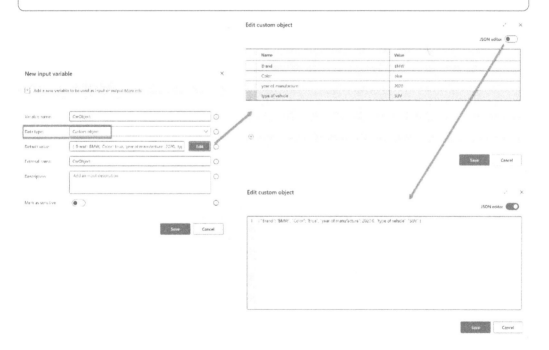

Figure 5.19 – Creating and editing a custom object

This data type is especially useful when we need to create a flow that communicates with some external web services that expect or return JSON objects. We will discover this later in this book.

To convert the custom object into a valid JSON string, we can make use of the **Convert custom object to JSON** action, which expects a custom object as input and produces a new variable containing the JSON string. The reverse operation is **Convert JSON to custom object**. Both actions are also in the **Variables** action group. We will look at examples of how these actions are used later in this book.

Text data type and corresponding actions

We have now learned about the complex data types that PAD contains out of the box. The **text** data type does not belong to this group of data types, but there is a large number of actions that deal only with this data type. Therefore, it makes sense from my point of view to mention it briefly here. These actions are in the **Text** action group and perform all the tasks you would expect from a text editor and can be very useful in flow design. These include **Split text**, **Get subtext**, **Crop text**, and many more. I would like to pay special attention to the **Recognize entities in text** action, which is capable of extracting certain entities from a piece of text using AI. I encourage you to try the corresponding example of Microsoft for this purpose. To do so, switch to the **Console** window and go to the **Examples** section. Under the **Text Manipulation** heading, you will find an entry called `Extract phone numbers and emails from text`. Edit this flow and look at the call to experience the power of AI while interacting with PAD for the first time.

The % notation

In all the examples we saw earlier, we needed to use the `%` notation, which is needed whenever we want to use a variable. A variable or an expression in which one or more variables occur must always be enclosed with the `%` character since it is precisely this part that is evaluated by PAD. Within the two `%` characters, simple arithmetic operations, logical comparisons, or expressions can be used together with variables or hard-coded values. Here are some examples:

- `%CarName%`: Returns the content of a variable called `CarName`.
- `%'CarName'%`: Please notice the quotes. This returns the hard-coded `CarName` value.
- `%Counter + 1%`: Returns the value of the `Counter` variable increased by 1.
- `%(Counter + 1) > 15%`: Makes use of parenthesis and returns a Boolean value of `True` if the `Counter` variable has a value of 15; otherwise, it returns False.
- `%NettoPrice + VAT%`: Returns the sum of the values of the `NettoPrice` and VAT variables.

> **Using the % sign as a simple character**
>
> Sometimes, we may want to use the `%` sign not in an expression but just as a normal character. In this case, it is possible to escape the character with another percentage sign (`%%`).

This notation is very straightforward to use and sometimes it is very handy to do a simple calculation right in the field where we would otherwise just place the variable itself. On top of that, within these expressions, it is also possible to make use of the variable's data type properties. This refers to those properties that are provided by the non-built-in data types. These include, for example, dates, lists, files, folders, Outlook messages, web browser instances, and more. For example, the **Get files in folder** action returns a list of files, each of them being of the **File** data type. We could iterate through this list and have the current item in a variable called `currentItem` to access properties such as the following:

- `% currentItem.FullName%` contains the full path to the file
- `% currentItem.Size%` is the size of the file in bytes

Please consult the *Further reading* section for a link to the official documentation of all data types and their properties.

Variables and the `%` notation are extremely important parts of creating flows. It is vital to become familiar with this concept and how variables can be created and manipulated. Now, let's look at how we can create UI elements to remotely control external applications.

Identifying and creating UI elements

In *Chapter 1*, we learned that PAD comes from the **robotic process automation** (**RPA**) area and that this involves remotely controlling local applications. It is those applications that supply input options, which can be used for flow processing. These input options are identified by PAD through so-called UI elements, such as buttons, lists, or other controls used by the application.

In this example, we want to control the built-in Windows calculator application. The easiest way to create UI elements is to use the recorder. Follow these steps:

1. Start the Windows calculator from your desktop and leave it open.

2. Create a new flow called *Windows Calculator Remote* or provide some other meaningful name.

3. Switch to the **UI Elements** pane on the far right-hand side and click the **Add UI element** button. This will start the unified recorder of PAD. You should now see the recorder window next to the Windows calculator. If you hover over the different numbers and buttons on the calculator, you will notice that these get highlighted with a red rectangle, which means that this will be identified by PAD as a UI element.

4. Now, press the *Control* key on your keyboard and left mouse click on one of the number buttons in the calculator window. This will add a UI element to the recorder:

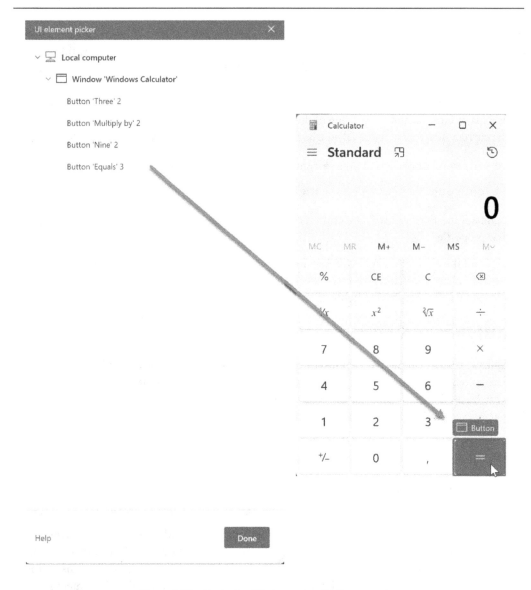

Figure 5.20 – Capturing UI elements with the recorder

As you can see, I used the button sequence 3, then * (multiply by), then 9, followed by = (the equals button), and ended by clicking inside the result text field.

You can now press the **Done** button in the recorder to transfer the captured elements to the flow designer. The UI elements pane should look like this:

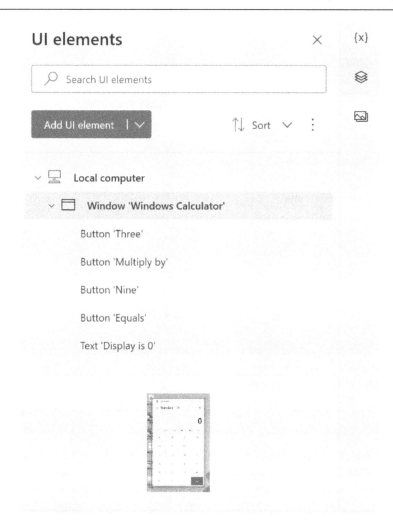

Figure 5.21 – UI elements hierarchy

Now that we have some UI elements, we can also demonstrate their use. PAD has dedicated actions for different UI elements. Since we have captured buttons, we could use the **Press button in window** action (the **UI automation | Form Filling** action group) to simulate the calculation by PAD. Just drag this action four times into the designer and select the right sequence of UI elements. It should look like this:

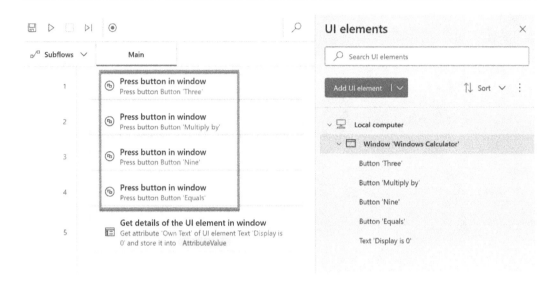

Figure 5.22 – Flow simulating a calculation

We can also use the **Get details of the UI element in window** action to extract the data of the result text field to use this somewhere else. Executing this flow will press the buttons according to the sequence in the flow magically. This is UI automation.

It is also possible to create UI elements for desktop icons or taskbar icons. In conjunction with the actions in the UI automation actions group, this provides us with an extensive repertoire of automation options.

However, not only Windows applications can be controlled remotely. PAD can also access and remotely control web applications and pages. The procedure is the same as for Windows applications. However, a corresponding browser extension must be installed for the browser being used so that PAD can recognize the web elements. Microsoft provides browser extensions for Google Chrome, Mozilla Firefox, and Microsoft Edge (see the *Further reading* section for more information).

Virtual desktops can also be included in automation. However, for UI elements to be recognized there, a small extension is also required – the so-called Power Automate agent for virtual desktops. Further information on this can be found in the last section of this chapter, as well as in the *Further reading* section.

Therefore, recognizing UI elements plays a very important role so that corresponding actions such as clicks or *button pressed* can be executed. PAD arranges each UI element hierarchically under a so-called selector and assigns names and properties to make the element unique in the corresponding application and the operating system. Double-clicking on a created UI element shows that selector and its properties, as shown in the following figure:

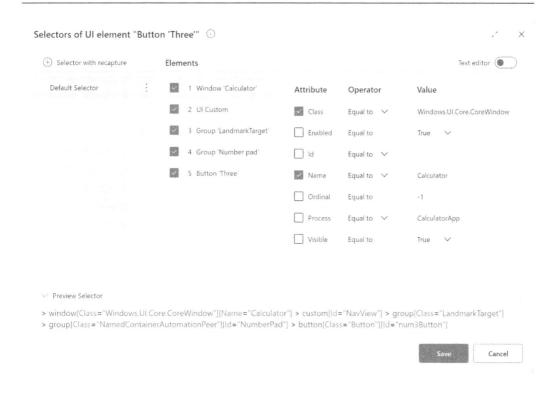

Figure 5.23 – The selector view for a Windows button

It is also possible to create a selector and disable an existing selector for an element. This can be especially helpful when testing because correctly identifying UI elements is crucial for automation.

We will look at more details and examples of this in *Chapter 6*.

How to use images with PAD

As we learned in the previous section, PAD can identify UI elements by using specific selectors. But there are situations where we need other mechanisms to perform certain actions, such as moving the mouse to a certain place and performing an action. This is where images come into play. We have the following actions that use images as input:

- **if image** checks whether an image is available on the screen
- **Wait for image** waits for an image to appear or disappear on the screen
- **Move mouse to image** moves the mouse to an image on the screen

Combined with **Send mouse click**, it is even possible to perform actions without knowing the UI element. This technique is also useful if we need to automate on virtual or remote desktops where there is no agent available, such as Linux or Unix. Any program on any system can be executed and controlled remotely via this technique. So, for example, if there is a business application that runs on a Linux system and does not have any API, it would be possible to automate this program with PAD as well.

> **Power Automate agent for virtual desktops**
>
> The computer running PAD can, in principle, also remotely control a **remote desktop** protocol **(RDP)** session. If the remote system is also a Windows system, the Power Automate agent for virtual desktops can be installed on that system. This makes it possible to use UI elements within the RDP session as well. This works with both Citrix and Microsoft Remote Desktop.
>
> For all other operating systems, you can only use PAD images to remote control.

To create or capture an image, we can use the third option on the far right-hand side of the flow designer window, which reveals the images area. Pressing the **Capture image** button enables you to click and drag a selection on the screen. This is supported by a magnifying glass so that it's as precise as possible. After releasing the mouse, you will be prompted to enter a name for the image. If we perform this procedure for any desired area on the screen, we will quickly have a library of images where we can work with the PAD actions.

The following basic example demonstrates the usage of images. Therefore, I am accessing a Linux session via the Microsoft Remote Desktop program. For this demo, I only want to launch the file explorer in that session. To do that, follow these steps:

1. Navigate to the **Activities** menu and left-click on it.
2. Wait for the taskbar to appear.
3. Press the button for the file explorer, as shown in the following figure:

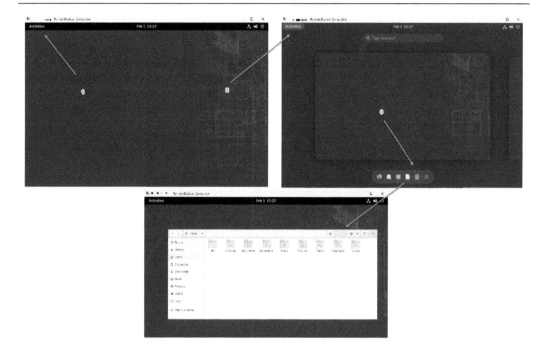

Figure 5.24 – Remotely controlling a Linux system

To be able to do that, I need to create two images in PAD via the procedure described previously: one for the **Activities** menu item and another one for the file explorer icon. The following figure shows the flow that needs to be created to execute this:

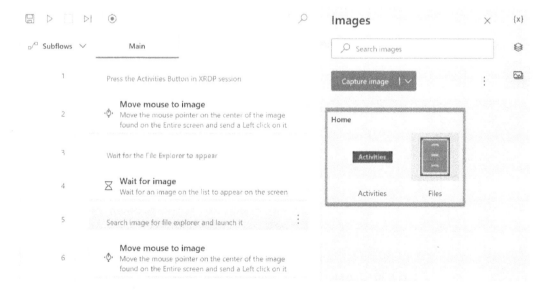

Figure 5.25 – Images in a flow

Let's take a look at the individual steps of the flow:

1. **Move mouse to image**: To use this action, an image needs to be provided – in this case, the **Activities** image. This action also allows us to directly send a mouse click after the move. In addition, I used the advanced image recognition algorithm to find the image on the screen:

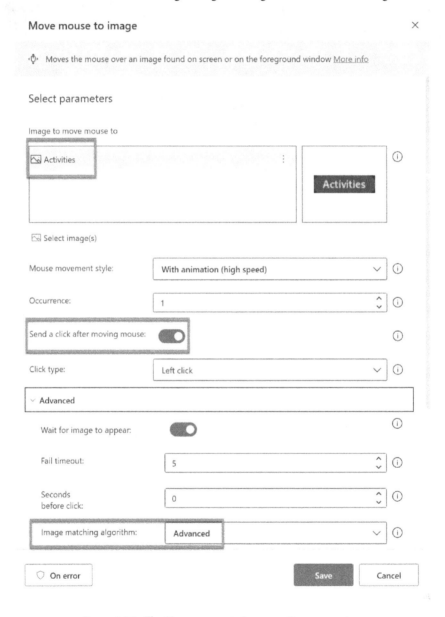

Figure 5.26 – The Move mouse to image action parameters

2. **Wait for image**: The next step is to move the mouse down to the file explorer image. But when the flow starts to run, this image is not yet on the screen. That's why we need to wait for the image to appear.

3. **Move mouse to image**: Here, the flow moves the mouse to the file explorer and launches it by clicking on the icon.

There are, of course, many more expansion stages that can be built with these capabilities. This small example shows the basic potential of PAD to remotely control applications that run on a completely different technology.

Summary

In this chapter, we learned about the input elements for PAD and their use cases. We now know what types of variables there are and how we can create and manipulate them. This enables us to communicate with other external applications, receive data, and also return it. UI elements and images help us identify and interact with the buttons and areas of applications on a local machine, enabling automated operation.

Now, we can use this knowledge to start implementing real business processes and addressing use cases with it using the additional actions of PAD. We'll start with the UI automation actions, which we'll study in the next chapter.

Further reading

To learn more about the topics that were covered in this chapter, take a look at the following resources:

- Using variables and the % notation: `https://learn.microsoft.com/en-us/power-automate/desktop-flows/variable-manipulation`

- Variables data type properties: `https://learn.microsoft.com/en-us/power-automate/desktop-flows/datatype-properties`

- Power Automate browser extensions: `https://learn.microsoft.com/en-us/power-automate/desktop-flows/install-browser-extensions`

- Automate on virtual desktops: `https://learn.microsoft.com/en-us/power-automate/desktop-flows/virtual-desktops`

- Automate on virtual desktops: `https://learn.microsoft.com/en-us/power-automate/desktop-flows/virtual-desktops?source=recommendations`

6

Actions for UI Automation

In the previous chapter, we learned about UI elements and how they can identify areas an application interacts with. In this chapter, we will highlight the possibilities of PAD that can be accomplished with actions for UI elements. This includes covering the following:

- Exploring the different UI actions and their purpose
- Using UI actions to enable data exchange (extract data and form filling) and automation between two applications

By the end of this chapter, you will understand what is required to identify the necessary buttons and UI elements and trigger them with UI actions to safely and reliably repeat the desired tasks and configure them in a flow.

Why do we need UI actions?

Over time, countless applications for Windows were created, of which a great many still exist to this day. Microsoft itself has created excellent options for tailoring your applications with tools such as Visual Basic and Visual Studio. However, many of these applications are still not accessible via an API and have to be operated manually – also known as the human interface.

As we will see later in this book, PAD can work directly with numerous and widely used standard applications such as Office and Outlook, but it would be impossible to cover the wealth of different applications with dedicated actions. And that is the reason why the actions for UI elements exist. The term **UI automation** refers to the general capability of PAD to work with any application window and therefore remote control any application and control element.

To enforce this very general approach, the actions in this action group are correspondingly general and can be divided into the following parts:

- Identify a window
- Interact with application controls

- Extract data from a window
- Fill in data in a window

Let's take a closer look at each of these actions before we dive into another example to leverage these actions.

Identify a window

The first thing we need to do is get the window of the application we want to control. We already know that there are several ways to start a program, such as by utilizing the *run application* action or using a dedicated action for a program such as Excel or Outlook. The following actions identify a window or make it accessible for PAD, especially in combination with other windows:

- **Get window/focus window**: This can be found in the **Windows** subgroup. You can get or focus a running window and return a title and/or Window handle.
- **Set window state/set window visibility**: This can be found in the **Windows** subgroup. You can set the state of a window to **restored**, **maximized**, or **minimized**, which hides a window or makes it visible again, respectively.
- **Move/resize/close window**: Also in the **Windows** subgroup, this action can move a window to a specific position of the desktop by providing *X* and *Y* coordinates, can change the size of a window to a specific width and height, or just close a window.
- **If window/wait for window/wait for window content**: This action executes actions in a conditional block, depending on whether a window is open/focused or not, waits until a window is opened/closed/becomes the focus/loses focus, and waits for a specific UI element that should be contained in a window.

These actions are especially helpful in situations where an application might have some latency when opening or during the operation. If PAD were to access a UI element that hasn't been displayed yet, these actions could be used to wait for the element to appear.

Interact with a window

PAD also needs to be able to remote control an application. This means clicking on a button or area or dragging and dropping an element. This is possible with the following actions, which are self-explanatory:

- Select a tab in a window
- Hover the mouse over a UI element in a window
- Click on a UI element in a window
- Drag and drop a UI element in a window
- Expand/collapse a tree node in a window

These actions work with UI elements that have been previously defined; for example, with the recorder. In the example for this chapter, we will make use of some of these actions to remote control the application.

Extract data from a window

We used some of these actions in one of our previous examples where we extracted the values of checkboxes in a Word document. There are more possibilities when it comes to extracting data from an application:

- **Get the details of window**: This action produces a variable containing properties of a window such as its title or source text. The desired property can be chosen in the actions dialog.

- **Get the details of a UI element in a window**: This action is the same as the previous one, except it's for a specific UI element and different attributes, such as your text and its location and size.

- **Get the selected checkbox in a window**: This action returns a variable with names of checkboxes in a checkbox group or the state of a specific checkbox.

- **Get the selected radio button in a window**: This action is the same as the previous one but only two values can be selected.

- **Extract data from a window**: This action returns a variable that can also directly be opened in Excel as a spreadsheet containing a single value, a list, or a table.

- **Take a screenshot of a UI element**: This action does what it says.

Fill in data in a window

Inserting data into an application is an important function to control applications remotely. The following actions can be used for this purpose (they are self-explanatory):

- Focus the text field in a window

- Populate the text field in a window

- Press a button in a window

- Select a radio button in a window

- Set the checkbox state in a window

- Set a drop-down list value in a window

Now that we know what actions exist to work within a UI, let's look at an example of how to incorporate these actions.

UI actions in action

In the example presented in this section, we want to establish a data exchange between two applications that can't do this: a Microsoft Access database and a Windows Desktop application. This scenario is illustrated in the following figure:

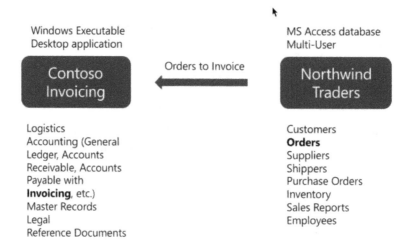

Figure 6.1 – Orders to Invoices structural view

In this scenario, a database application has been created with MS Access. This application may be used by multiple users through network sharing and contains information about customers and suppliers. The users can also store orders from the customers in this database, but the invoices to customers need to be processed via another application for some reason. This application is a Windows executable program that is running on the desktop of a user.

Microsoft Access database as a source system

The database we are using in this example is a sample database provided by Microsoft. Starting MS Access from the program taskbar in Windows provides also provides a lot of sample databases from which the *Northwind* database is one of them. We can use the search function to look for this example, double-click it, and store it locally in our Documents folder.

This database contains a lot of functionality, starting with the possibility to have different user roles and nice dialogs to input data (see the following figure):

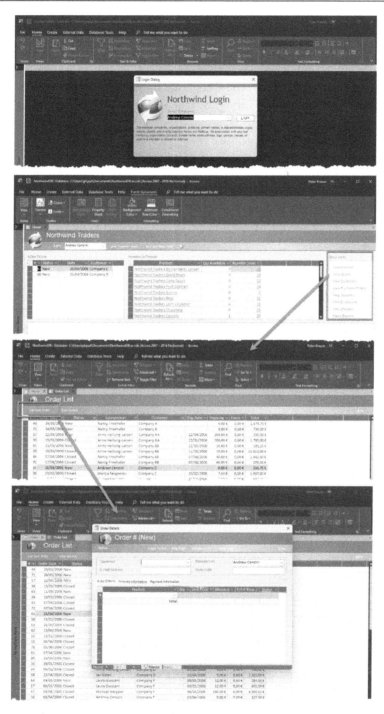

Figure 6.2 – Northwind Access application

In our example, we assume that multiple users will add their orders to the database. But to create invoices out of these, the data must be transferred to the invoicing application. Neither application has built-in capabilities to transport the data elsewhere.

Having an Access database as one of the central applications is a very common scenario in smaller and even sometimes bigger companies. This is also true for the second application we are going to use in our scenario.

Windows executable as the target system

We are going to use a sample application provided by Microsoft called **Contoso Invoicing** as the target system. This is just an example of an ordinary Windows application with specific characteristics such as a navigation tree, a menu, a toolbar, and some forms to view and capture data. Please refer to the *Further reading* section for more details on how to install this application, which looks like this:

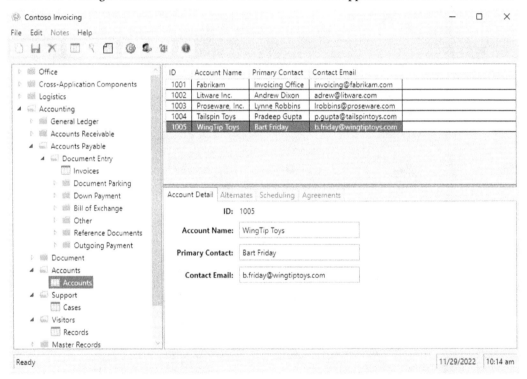

Figure 6.3 – The "Contoso Invoicing" application

A desktop icon is also created during the installation, which we will make use of shortly. One additional thing to mention here is that an Excel workbook is also provided by the installation, which contains all the data for this application. This is located in the standard user's documents folder under `Contoso Invoicing - 1.0.15.0`. We can modify this Excel workbook to our needs and, for example, remove our sample data again.

To be able to create invoices, we must call up the **Invoices** menu item in the navigation tree, as shown in the preceding screenshot, and then enter the data. This outlines a thoroughly typical example of the setup in small and medium-sized companies.

Deconstructing the workflow

This is a non-trivial flow, so it is best practice to cut this into smaller pieces. Those smaller flows will be easier to maintain and test since with the interaction of UI elements, a lot of actions will always be used:

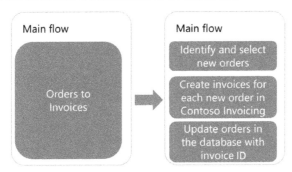

Figure 6.4 – Deconstructing the Orders to Invoices workflow

1. To identify which new invoices need to be created, we first need to identify these in the Northwind database and store the orders in a data table. Next, we must iterate through this data table and create new invoices in the Contoso Invoicing application. Invoice IDs get generated for each new invoice that we can also save in the data table.

2. The last routine takes these new invoice IDs and updates the database with that information.

Hint: All three routines are separated into the following subflows. The reason is to identify issues related to a specific action and improve the maintenance of flows:

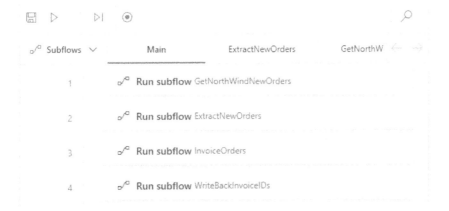

Figure 6.5 – Subflows for Orders to Invoices

To create the whole workflow, we will need to apply all the knowledge that has been discussed throughout this book, which is flow control, variables, and UI elements. Let's take a look at the different subflows we can use to accomplish these tasks.

The GetNorthWindNewOrders subflow

In this subflow, we want to start the database application and, in the end, get a list of all the new orders that have arrived in the database.

To remote control the application, we need to define or capture the corresponding UI elements that we want to use to achieve the desired result. In the previous chapter, we learned how to do that. In addition, some variables are defined. This is the detailed flow to get new orders:

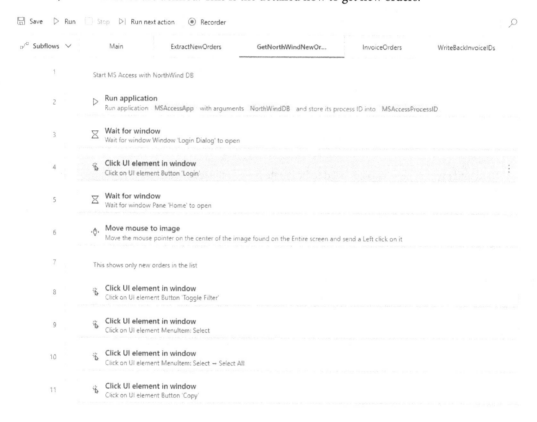

Figure 6.6 – Subflow to extract new orders

These are some comments on the different steps (line numbers).

Step 2 starts Microsoft Access with the Northwind Database. I used two variables here. MSAccessApp contains the full path to the executable of Microsoft Access, while NorthWindDB contains the path to the database itself.

In *step 3*, we use the first UI action, called **Wait for window**. Old programs or computers with less power (processor, RAM, and so on) might have some latency when starting a program. This action will pause the flow until the given window appears. In this case, the flow waits for the login screen to appear. The login window was captured as a UI element previously.

Step 4 triggers the click on the login button. As shown in *Figure 6.2*, the database application first wants a user to log in. Here, we just take the first user and press the login button.

Now we should wait for the main application window to appear. That's why we use the **Wait for window** action again in *step 5*.

In *step 6*, I used an image to be able to address the **view orders** link on the right-hand side of the application. The recorder was not able to detect this as a separate UI element, so I had to use the image functionality of PAD to move the mouse to this link and click on it. This is a good example showing that even those cases can be addressed by PAD.

In *steps 8* to *11*, I am using MS Access ribbon buttons to filter the list of orders so that it only contains new orders, selecting all of them, and copying them to the clipboard. This allows you to parse the content in the next subflow.

Please notice that we are not closing the application at this point.

The ExtractNewOrders subflow

At this point, we only have some text in the clipboard. But to be able to iterate through the list of new orders, we somehow need to convert this text into a data table. The following figure shows the flow and how to accomplish this:

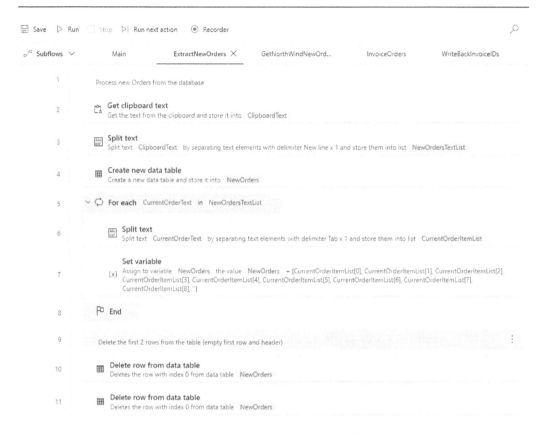

Figure 6.7 – Generating a data table from clipboard text

After we stored the clipboard text in *step 1* into a text variable by using **Get clipboard text** in the **Clipboard** actions group, we can now split this text into a text list (*step 3*).

Next, we create a data table variable with the right columns. These are all columns from the MS Access orders table, plus one additional column for the invoice ID, which gets generated later by the Contoso Invoicing application:

Figure 6.8 – The internal order data table definition

In *step 5*, we start to iterate through the text list, and for each entry in that list, we need to split the text again to get the single values for each column. These get stored in the data table in *step 7*.

Finally, we need to clean up the data table and delete the first two rows (row 1 was empty and created while defining the data table, while row 2 contains the column header information from the clipboard text).

Now, we have a data table containing all the new orders that we can work with.

The InvoiceOrders subflow

This subflow is responsible for creating the invoices in the Contoso Invoicing application. Here, we also need to identify the corresponding UI elements first, such as the **Invoice** tree node, the *new* (add button icon) and *save* (add button icon) buttons, and so on. The following figure shows the required UI elements:

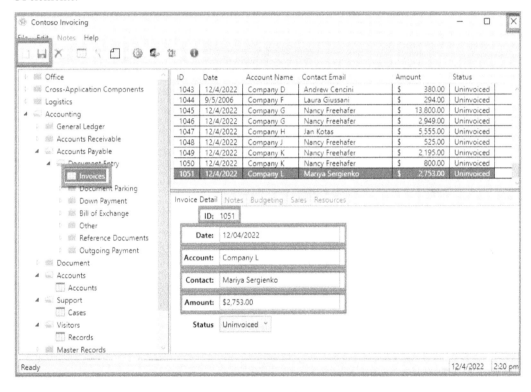

Picture 6.9 – UI elements in Contoso Invoicing

As we can see, we need more UI actions to fill in and also extract the data. The following figure shows the flow that does the required steps for us:

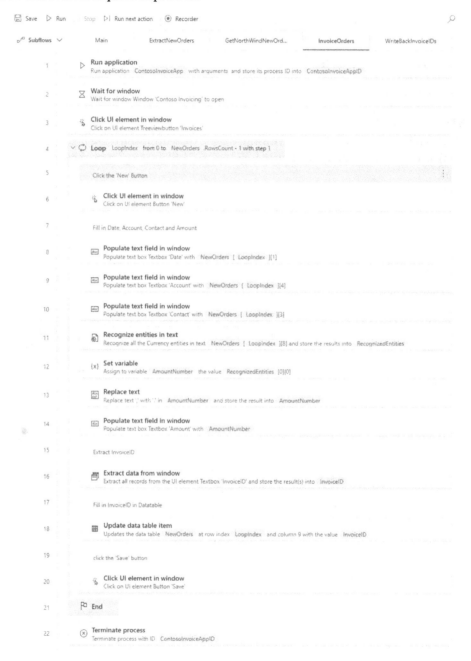

Figure 6.10 – Invoicing orders into Contoso Invoicing

Again, let's walk through the steps to understand the details:

1. *Step 1* launches the application.

2. *Step 2* waits for the main application window to appear.

3. *Step 3* clicks the **Invoices** tree node. We are ready to insert invoices now.

4. *Step 4* starts a loop and iterates through the data table for new orders.

Notice that I chose to use the indexed version for iteration. This makes it easier at the end to fill in the data for the generated invoice ID because I can reference the current iteration step via a loop index.

In the loop, we do the same for each of the data rows in the table:

1. Click the **New** button (*step 6*).

2. Fill in the data for the date, account, contact, and amount (*steps 8 to 14*). Here, we use the **Populate text field** action from the **Form filling** actions subgroup. There is also one special thing we need to take care of. Because we extracted the data from the database as text, the currency field contains a value that the Invoicing app does not understand. A conversion is needed to put in the right value. To convert the amount value, we could use the **Recognize entities in text** action. In this case, an entity is something such as an email, an IP address, or a currency value. This action uses artificial intelligence and analyzes the given text, along with the rules on what to detect. The action returns a data table containing all the detected values. It can be configured like this:

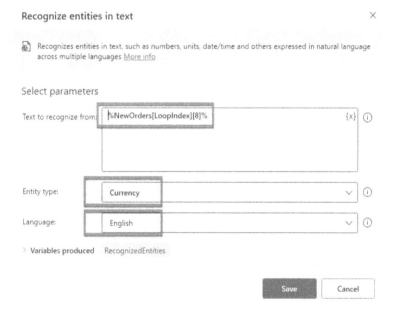

Figure 6.11 – The Recognize entities in text configuration

3. Extract data for the generated invoice ID and store it in the last element of the data row. Here, we make use of the **Extract data from window** action to capture the generated invoice ID.

4. *Step 18* updates the orders data table with the extracted invoice ID.

5. In *step 20*, we click the **Save** button.

At this stage, we have established the communication between the two applications and inserted new orders as invoices into the Contoso Invoicing app. In addition, we now know all the invoice IDs for each order so that we can update the MS Access database itself for reference. Let's see what we need to do here.

The WriteBackInvoiceIDs subflow

Finally, we want to update the MS Access database so that users can see that invoices have been created. Now, this could be facilitated in many ways. Unfortunately, the database itself does not provide us with an easy way to mark an order to be invoiced. That's why we can just choose an easy way to pass this information back into the database by updating the **Notes** field in each order. We will do that by dynamically creating a SQL statement that contains the order ID, as well as the invoice ID (both are in our data table). This statement could be like this, for example:

```
UPDATE Orders SET Notes = "Invoice ID: 1159" WHERE [Order ID] = 59
```

The flow that does this for us is displayed in the following figure:

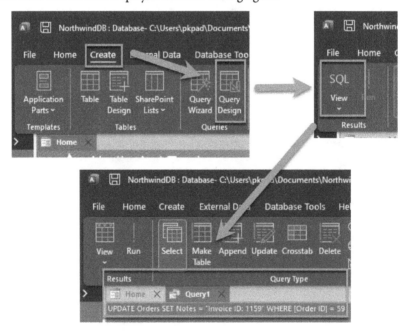

Figure 6.12 – Navigating to the SQL pane

To be able to paste the query text into MS Access, we need PAD to navigate to a place where we can work with queries. You can do this by clicking the **Create** ribbon, then the **Query Design** button. After that, we need to switch to the SQL view to be able to enter the query there. This is what the flow does in *steps 3* to *6*.

Please notice that we are running through this routine for every record in the orders data table. In *step 8*, we create the updated SQL statement and again use the form-filling UI action called **Populate text field in window**.

There is a **Run** button in MS Access that can be used to execute the query. In *step 9*, we instruct PAD to press this button. The default behavior of MS Access is to make a dialog appear, after which we have to confirm the action. In *step 12*, we tell PAD to do so.

Next, we want to close the query window and start over again. In *step 13*, we press the button to close the query window; again, we need to let PAD confirm that we do not want to store the query. This marks the end of the loop so that we can start over again:

Figure 6.13 – Updating MS Access with invoice information

This whole procedure seems a bit cumbersome at first glance, but it reveals the following:

- PAD is fully capable of interacting with applications and confirming dialogs

- It does not matter whether it is an application of Microsoft or not, because the concept of UI elements is universal

- PAD can extract data from other applications and use them with UI elements to create further actions

- If the task presented here has to be performed once or even several times a day (which could well be the case), then the integration and automation shown here pay off several times more

In addition to these important points, I would like to make the following comments:

- As already mentioned, there is also a billing function in the database. However, before an invoice can be created, further conditions must be fulfilled, which I have ignored for the benefit of this example.

- When handing over the amount, we have not performed any currency conversion for the sake of simplicity.

- All the UI elements that we used were previously inserted into PAD via the recorder. *Chapter 5* described how this can be achieved in detail.

In this section, we learned how PAD bridges the gap between two different applications that need to exchange data with each other. By performing data extraction, we can read information from one application, convert it into another format, and insert it back into a second application by using form-filling UI actions.

Summary

In this chapter, we learned how PAD allows two applications to exchange data that would otherwise have no connection. By defining UI elements and using UI actions, any application can be remotely controlled and thus integrated into a larger automation context. Here, PAD acts as a high-level entity that allows the application landscapes to communicate with each other; otherwise, no data exchange would be possible at all unless you were to do it manually.

In the next chapter, we will look at other actions that can be used to manage and remotely control a computer, its files, and its services.

Further reading

The following walkthrough, *Microsoft 'RPA in a Day' Training Material Update*, provides a full course with different modules that also leverage the sample application used in this chapter. The article contains a link for you to download the material. In the `Lab data packages` folder, there is another folder called `Prerequisites…` that contains the installer for the Windows application: `https://powerautomate.microsoft.com/en-us/blog/announcing-microsoft-rpa-in-a-day-version-2/`.

UI automation reference: `https://learn.microsoft.com/en-us/power-automate/desktop-flows/actions-reference/uiautomation`

7

Automate Your Desktop and Workstation

In this chapter, we will take a look at the huge number of actions that that are part of system management and that also have high practical relevance. Some of these actions that we will cover in this chapter are the following:

- Actions for workstation management, including scripting
- Files and folders
- Computer peripherals such as the mouse and keyboard, printers, and clipboard

We will dedicate the first part of the chapter to the action groups that make the management of a system possible before exploring the possibilities in the area of folder and file management.

The last part deals with the automatic operation of the peripheral devices of a computer. We will create a PowerPoint slide deck to document the status of a local workstation. By the end of this chapter, we will have learned that **Power Automate desktop (PAD)** can manage all important aspects of the local computer and that this provides a powerful tool for folder and file management.

Technical requirements

Operating system (OS) processes and Windows services are crucial components, and incorrect manipulation could have an impact on the stability and reliability of the Windows OS. That's why administrator permissions are required to work with these components.

You can find the technical requirements at the following link: `https://learn.microsoft.com/en-us/power-automate/desktop-flows/requirements`.

Windows and desktop management

The mass management of workstations in small and large networks can be supported by numerous tools today, such as Active Directory group policies, Microsoft Intune, Microsoft System Center, and so on. Typical administration tasks in the administration area are as follows:

- Installing and setting up the OS

- Creating, configuring, and authorizing user access

- Installing additional standard software for users, such as Office or other applications

- Regular maintenance and health checks are performed by executing commands or scripts and monitoring and managing processes and Windows services

As mentioned previously, these tasks can be partly covered by other mechanisms, such as Active Directory group policies, Microsoft Intune, Microsoft System Center functionality, or other third-party management software. Typically, however, such tools require greater effort to procure and administer, and the software is aimed at professional IT administrators in larger environments. These tools also require local agent software (a Windows service that is installed) that frequently contacts the central administration instance for new updates, which can subsequently be deployed.

Looking at the capabilities of PAD in the following section, it appears to be a useful tool in this area. To clarify the positioning of the different approaches to remote management software and PAD, we can look at the following table:

Feature	Remote Management Software	Power Automate Desktop
General approach	Consistently manage desktops and OSes by providing a set of software packages and monitoring devices and OS security	Executes desktop actions that cannot be performed automatically and remotely controls the operation of software by simulating a user
Provide management of desktops	Centrally, with remote agents to be installed on each machine	Distributed – a desktop flow is created on one machine and can be shared afterward
Install software packages	Automatically after central configuration	The creation of a flow is required for each package to be installed
Ability to operate software	Not available	The creation of a flow is required for operating software

Table 7.1 – Comparison of Remote Management Software and PAD

As we can see in the preceding table, PAD is not intended to cover centralized network management, although we find several action groups that could be used to create a flow to perform administrative tasks, some of which are not possible or not available with centralized network management.

On the other hand, it is not possible with such management software to map individual operations of one or more software products on a desktop.

Therefore, the following use cases arise for PAD in this environment:

- There is a need for the management of workstations, but no management software is available
- A task must be executed on a local computer that cannot be mapped by such management software
- A task that must be executed is not available or not possible with such management software
- This task must be executed more frequently and/or on multiple workstations

So, let's take a look at what PAD brings to the table.

Controlling processes and services

The **System** action group contains many actions that can be used to control the processes of the local computer as well as local environment variables.

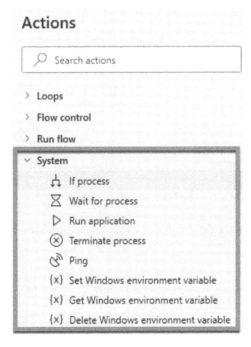

Figure 7.1 – System actions in PAD

A process in Windows represents a program that has been started. Such a process has both a name and a unique ID. We can view all started processes in Windows by executing the `tasklist` command on the command line. So, if we run Microsoft Word or any other program, the corresponding process will be started for it. Typically, users interact with processes that are started for a specific program. PAD has a **Run application** action that also does this. We have already used this action multiple times in our previous examples. We could use this action to start any application on the desktop or open a document that is associated with a specific application by providing some command-line arguments, as we see in the following screenshot:

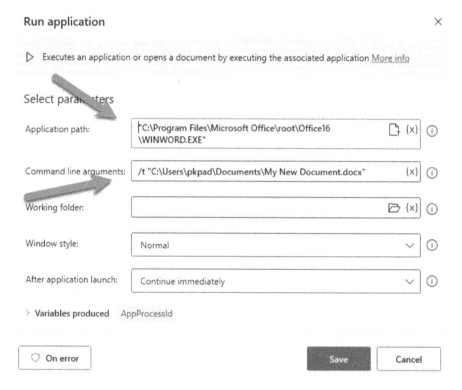

Figure 7.2 – Run application action parameters

The **Application path** parameter determines the executable program, including the full path to the application.

In the **Command line arguments** parameter, we can enter all additional parameters that the application knows and that we want to bring in. In the preceding example, we used the command-line switch for opening a document.

Tip for finding out the complete path of an application

If you want to find out what the full path to an application is and there only is an entry in the Windows start menu, do the following:

1. Identify the application by typing its name, then right-click on the icon, and choose **Open file location**. This will open Windows Explorer to the shortcut of the application

2. Now right-click on the shortcut and choose **Open file location** again. This will locate the executable program in Windows Explorer.

3. Finally, right-click on the application and choose **Copy as path** or *Ctrl* + *Shift* + *C* and paste this value into the **Application path** parameter field.

If you want to use a specific working folder for the application, you can specify this in the corresponding parameter. **Window style** determines whether the application window will be displayed as normal, maximized, minimized, or hidden.

Some applications take some time before they reveal an application window. This means the process has already started, but the **user interface** (**UI**) is not yet displayed. Another reason for this might be the performance of the local computer. But if we need this application window for further processing, we could specify what to do when the application launches: continue immediately or wait for the application to load or complete.

When this action has been executed, the process ID of the application will be stored in the corresponding variable. This variable can be used later to close the program by terminating the process.

To compensate for the latency of rather poorly equipped computers, it is helpful if PAD can wait for the availability of an application and thus a process or check whether a process is already running (or no longer running) in order to start the next resource-hungry application if necessary. This is where the **If process** and **Wait for process** actions come into play, doing exactly what they say they do, and for which they need a process name:

- **If process**: This checks whether a given process name is or isn't running

- **Wait for process**: This suspends the execution of the flow

Next, let's discuss a scenario where these actions might be very useful:

- We assume that there is a very heavy, large desktop application that we can only afford to run once on the computer because of limitations of CPU and RAM.

- When we launch the application manually, it takes some time before the UI gets loaded and the application is operational. Here, we could use the **Run application** action and set the **After application launch** parameter to **Wait for the application to complete**.

- We have to make sure that there is only one instance of this application. Otherwise, the computer won't be responsive anymore. The application also sometimes crashes, so we want to make sure that any orphaned process gets terminated before we start a new one. The **If process** action could help here to check whether there is still another process running for the application, and we would use the **Terminate process** action to shut this process down.

- We need some operations to run in that application very frequently, so we want to use PAD to automate these tasks. In this step, we would design our flow and create UI elements and actions to perform the tasks that we need.

- After our work on the application, we want to ensure that the application is closed properly and completely. We could make use of the **Wait for process** action here and check whether the process is terminated. We could also take care of any timeouts for this action and retry terminating the process.

> **Tip to avoid many configurations of different devices**
>
> Introducing a device policy taken care of by the IT department will avoid fragmented hardware across your organization.

The **System** action group also has additional actions, and we have summarized them as follows:

- **If process/Wait for process**: Previously discussed.

- **Run application/Terminate process**: Previously discussed.

- **Ping**: This action sends a network message to a remote computer and checks whether the remote machine is accessible.

- **Set/Get/Delete Windows environment variable**: These environment variables can store data and values in a central place used by the OS and all other installed programs. There are standard environment variables such as **WINDIR** (the installation location of the Windows OS) or **PATH** (a list of directories that contain important programs that are directly accessible via the command line). These corresponding actions can manage these entries within a PAD flow.

In addition to running applications (processes) with a UI, an OS also has the concept of background services that don't interact with a user. These services are also processes because once they are started, they appear in the list of active processes commonly wrapped by an application called svchost. exe (host application for a service). But they need special handling, which we have dedicated PAD actions for, located in the **Windows services** action group (see the following screenshot):

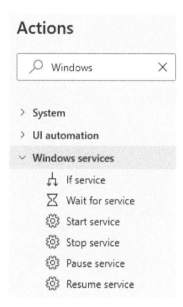

Figure 7.3 – The Windows services action group

The concept, however, is generally the same as with applications. There are the following actions:

- To start and stop a service

- Pause and resume a service

- Check whether a service is running or not and wait for a service to start or to stop

Sometimes, applications require a specific service to run or even install their own services. Large software installations and server products typically come with a set of services that they install during installation and that they rely upon. Services also can run into errors and need to be restarted. In such cases, a PAD flow can handle this by using the corresponding actions for services. The following simple example shows how these actions can be used:

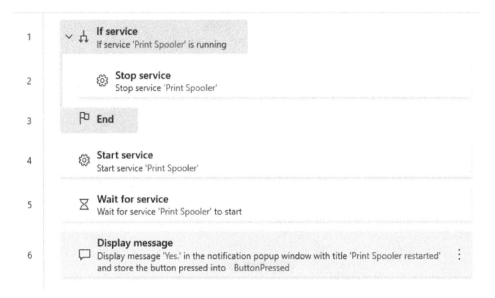

Figure 7.4 – Restarting the Print Spooler service with PAD

This desktop flow performs the following steps:

1. First, it checks whether the Windows Service **Print Spooler** is running.

2. If so, it stops the service by using the **Stop service** action.

3. After that, it starts the same service again.

4. In line 5, the flow waits for the service to start and suspends the execution until then.

5. Finally, a message box is displayed with a success message.

Because services can fundamentally affect the functionality of the OS, normal users without administration permissions are typically not allowed to manage services on a workstation. Also, PAD needs elevated permissions for the execution of this flow. Running this flow with the normal standard user permissions will result in an **Access Denied** error message.

> **Starting PAD with elevated permissions**
>
> To successfully execute the preceding flow, PAD must be executed with higher authorization. To do this, PAD must first be closed completely. The background process in the system tray must also be closed. Then, PAD can be started from the Windows menu. To do this, right-click on the program icon and select the **Run as administrator** option. Afterward, PAD can be used normally, and the flow runs without errors.

Another topic worth mentioning is a helpful feature of the actions for **Processes** and **Services**, which makes work easier when creating desktop flows. For both **Processes** and **Services**, the selection of the relevant element is made considerably easier by prefilling a drop-down list. The following screenshot shows the prefilled list of services that can be manipulated by PAD:

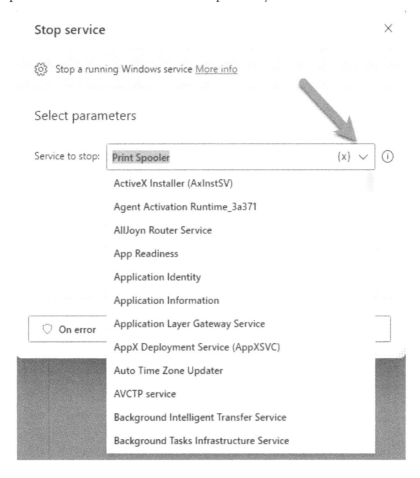

Figure 7.5 – Prefilled list of Windows services

It is, therefore, possible to select the element to be processed from a list of services/processes. We do not need to know directly what the process in question is called for Microsoft Word or any other program. To do this, we can simply start the program once and then select the correct one from the list of processes and use it afterward in our flow. And, of course, it is still possible to use a variable in these parameters.

Command-line sessions and scripting

In this section, we will look at the possibilities of using PAD to execute your own commands via the Windows command line (CMD) and scripting statements, intercepting the results and processing them further. To do this, we will look at the following action groups:

- **CMD session**: This performs the following actions: **Open CMD session/Close CMD session** as well as **Read from CMD session/Write to CMD session/Wait for text on a CMD session**

- **Scripting**: Run a DOS command/VBScript/JavaScript/PowerShell Script/Python script

Every OS has a command line (sometimes also referred to as a **console**), and sometimes that is the only form of access to the system. Therefore, the possibilities are very extensive, and any action that would otherwise be done with a UI can be performed and much more. The Windows CMD is the standard tool for issuing commands at this point. The command line can, of course, also be started manually via the Windows program menu, the Windows key + R shortcut, and the cmd command. The following flow uses all actions that we find in the **CMD session** actions group:

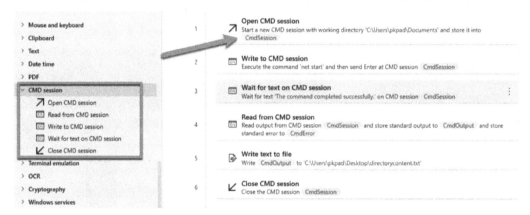

Figure 7.6 – Using the command-line actions in PAD

This flow has the following steps:

1. **Open CMD session**: This is always the first action when working with the CMD because this action creates the CmdSession variable, which is used by the subsequent actions to issue the command. It is also possible to set a specific working folder for the session in the parameters.

2. **Write to CMD session**: Here, the command specified in the corresponding parameter is passed to the command line and executed. In our example, we use the net start command, which returns a list of all started services on the workstation.

3. **Wait for text on CMD session**: If a command takes some time to execute, this action pauses the further execution of the flow and waits for specific text to appear on the command line. In our example, the command returns the The command completed successfully

text, which we use here. Other commands may return other results or even nothing. Please note that this action blocks the flow until the given text is returned. It may be necessary to also set the timeout for this action and handle the error.

4. **Read from CMD session**: This action takes the output of the command and stores it in a CmdOutput variable. Apart from the regular return value, there is also an error output for which a variable is also created and can be processed further.

5. **Write text to file**: This action does not belong to the **CMD session** action group but to the **File** action group, which is discussed later in this chapter. The action takes a variable and writes the content to a text file which can be defined in the parameters. We are creating a file saved on the desktop of the user workstation.

6. **Close CMD session**: The previously opened session is closed.

We have created a text file containing all the started services in this flow, and we could do some more investigation or send this now via email to a central maintenance mailbox for further processing.

Another way of executing a specific set of commands is by using the scripting functionality in PAD. The **Scripting** action group provides five scripting engines that work as illustrated in the following diagram:

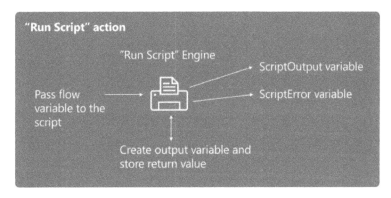

Figure 7.7 – The basic function of a Run Script action

Let's take a closer look at the concept depicted in the preceding diagram:

1. The first thing to notice is that there is a **Run Script** action mentioned, and this allows us to use all of the representative engines available in PAD, which include DOS command, VBScript, JavaScript, PowerShell Script, and Python Script. Scripting can be very useful for smaller blocks of code for certain functionalities that would otherwise be very cumbersome or not available at all. The different scripting engines each have their own strengths and give us the option to select the scripting language we already know. It is also conceivable that certain functionalities have already been implemented elsewhere with a scripting language so that this implementation can be used again here.

2. Now, to integrate a script into a PAD flow, we need to pass data and values to the script and play the result back. We can do this very easily using the % notation we already learned about in the chapter on variables.

3. The next step is implementing the actual functionality with the respective scripting language. At the end of this implementation, another variable should be defined, which contains the return value, and which is written back via the standard output. This value is then available in the variable that was defined via the action for calling the scripting block. Let's look at a little example here:

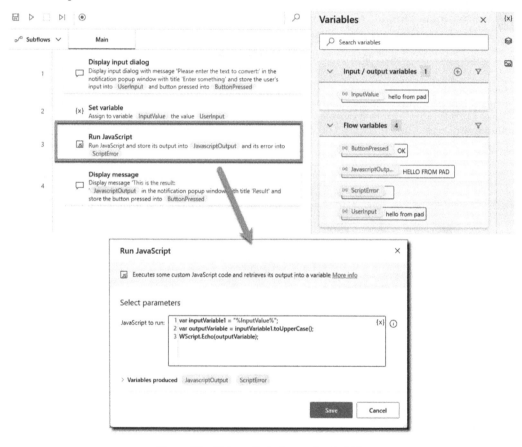

Figure 7.8 – Using a Scripting action in PAD

This flow shows the general concept of using **Scripting** actions in a PAD flow by implementing some simple scripting. In this example, we want to ask the user for some text that should be converted:

1. **Display input dialog**: This presents a message box to the user to enter some text.

2. **Set variable**: Here, we set the `InputValue` input variable to the text that has been entered by the user.

3. **Run JavaScript**: We use the **Run JavaScript** action as an example here. In the script, an `inputVariable1` internal variable is set to the value of the `InputValue` flow variable. It is also possible to transfer multiple variables into the script. The script then defines an `outputVariable` output variable and assigns the value of the input variable in capital letters by calling the `toUpperCase()` JavaScript function and writes this value to the standard output via `Wscript.Echo(outputVariable)`. With this, that value is subsequently available in the `JavascriptOutput` flow output variable.

4. **Display message**: The last step in the flow just displays the result of the processing in JavaScript.

I would like to add the following comments to the previous example:

- The flow is just an example to show the basic functionality when working with **Scripting** actions. The result of this process could also be achieved by using built-in functions, of course.

- The underlying Scripting hosts have a specific version that also defines which Scripting features are available. The **Windows Script Host** (**WScript**) had version number 5.812 (just output `WScript.Version` in a variable). The Python host allows Python 2 scripting, and the PowerShell version is 5.1.22621. Since we operate on a Windows machine with PAD, it is also possible to instantiate an ActiveX object, for example, an instance of Microsoft Word, and do everything possible with this kind of technology (see `https://learn.microsoft.com/en-us/power-automate/desktop-flows/how-to/extract-text-word-document`).

- Each **Scripting** action has a switch to separate the output of the script from a potential script error. It is recommended to enable this switch and have a dedicated variable for the script error. This makes it easier to debug the script because the second variable directly contains the error message. For the sake of simplicity, we did not implement any error handling in the flow, which is also recommended to do in a real-world scenario.

- You probably noticed that there is also a **Run DOS command** action in this action group. The difference from **CMD session** is that there we can issue one command after the other and check the result, while with **Run DOS command**, we define a script with several commands and run them as one block.

In this section, we learned that PAD offers a lot of possibilities for configuring a local computer. In the next section, we will look at the possibilities for handling folders and files.

Handling files and folders on your desktop

Working with folders and files can also be automated with PAD, for example, to perform copy or sort operations or to file files in specific folder structures. We have already learned about some of these actions in our opening example in this book. First, let's take a look at the actions that have to do with folders, which include the following:

- `If folder exists`: This starts a conditional block and checks whether a folder exists or not

- `Get files in folder`/`Get subfolders in folder`: This creates a list of files or folders that are contained in a specific folder

- `Create`/`delete`/`empty`/`copy`/`move`/`rename` `folder`: These perform the specific action on a given folder

- `Get` `special` `folder`: This retrieves the path of a special folder in Windows such as **Desktop**, **Internet Cache**, and so on

As we see here, no wishes remain unfulfilled. In combination with the following actions for files, these capabilities are further completed:

- `If` `file` `exists`/`Wait` `for` `file`: These can be used to check whether a file exists/ does not exist or whether a file is created/deleted.

- `Copy`/`Move`/`Delete`/`Rename` `files`: Operations for one or more files.

- `Read`/`Write` `text` `to` `file`/`CSV` `file`: Has been used in the previous example (CMD sessions).

- `Get` `file` `path`: This action produces several variables containing the root path, the directory, the file name with and without extension, and the extension itself.

- `Get` `temporary` `file`: Unlike the name suggests, a temporary empty file is created here that can be used for further purposes.

- `Convert` `file` `to` `Base64` or back/t4o binary or back: File conversion might be necessary in some cases to transfer files over a specific protocol and back to make it readable again. Base64 encoding, for example, encodes binary data as printable text.

Another action group worth mentioning in this context is **Compression**, which deals with one or more files that need to be compressed or decompressed. This is the same functionality that is also available when we use Windows Explorer and right-click any file and choose **Compress to Zip**.

Now let's look at an example of how these actions can be used to create a flow to copy daily backups of specific files to a network share. Conceptually, the flow should perform the following actions:

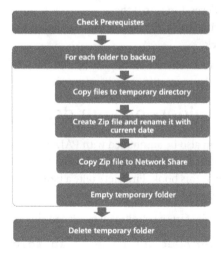

Figure 7.9 – Backup data with a desktop flow

First, let's take a look at the prerequisites that need to be checked before the actual backup is performed. In order to be flexible with the files to be backed up, it is possible to define all folders to be backed up in a text file and read them in first. Afterward, we can still check whether all necessary temporary folders exist.

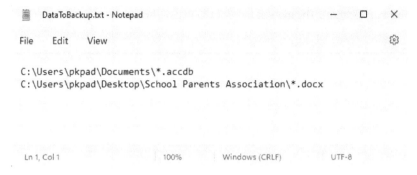

Figure 7.10 – Backup control text file

These preparations can easily be outsourced to a subflow. This would look like this:

> **Tip**
> The reason for creating a subflow is just because of maintenance and performance. It will technically work in one flow.

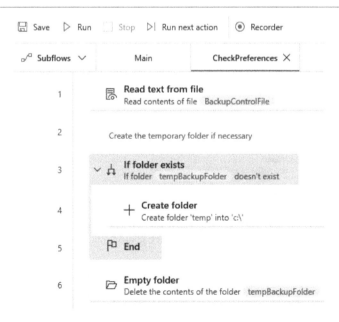

Figure 7.11 – The backup prerequisites subflow

Please note that corresponding input variables have been created to store the values for the backup control text file (see Figure 7.11) and the temporary backup folder.

The main flow now starts with a check of whether the backup control file exists and then uses the different file and folder actions to collect the data, create and rename the Zip file, and move each zip file to a given backup location. The main flow could look like this:

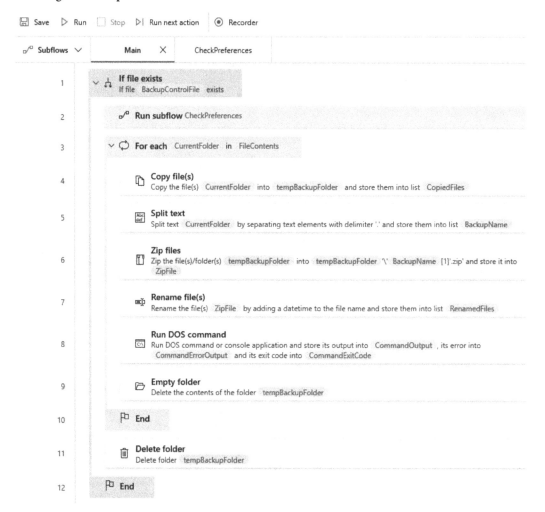

Figure 7.12 – The main backup flow

The heart of this flow is the `for each` loop, which takes each line from the backup control file and copies all files in that folder to a temporary location, creates a unique file name (line 5), compresses all files into an archive, renames the archive, moves the archive to a network share via a DOS command, and empties the temporary folder to start over.

As you can see in the preceding flow example, we used a lot of the actions included in the action groups we have discussed in this section. Please note that all these actions have many additional parameters. For example, in the Zip action, you can also assign a password for the archive. Unfortunately, we cannot discuss every action in detail; that would go beyond the scope of the book. However, it is important to understand what basic functionalities are hidden in the action groups so that you know where to look to master a particular task.

In the last section, we will explore the actions to interact with a workstation and address the peripherals.

Operating computer peripherals

The actions presented in this section complete the capabilities of PAD to manage workstations by simulating an operation or, of course, to cover other use cases. The following components can be controlled and can be found as action groups:

- The workstation itself and connected printers
- The mouse and keyboard
- The clipboard

With these action groups, it is possible to simulate complete user sessions by moving the mouse or pressing keystrokes.

This can also be helpful, for example, when none of the previously discussed options (UI elements or images) are available to communicate with an application. Maybe we are not able to capture UI elements, but then we can still navigate the mouse to specific positions and send clicks or even keys.

However, at this point, we have to be very precise, and this is for the following reason. We will see in this example that the **Send mouse click** action works with the X and Y coordinates of the screen. This means that the mouse is guided exactly to these coordinates, and then the click is triggered. This will work until the screen resolution of the local machine is eventually changed or the flow is run on another machine with a different resolution. The same coordinates then lead to a completely different location, so the flow no longer works properly. The relationship is shown in the following diagram:

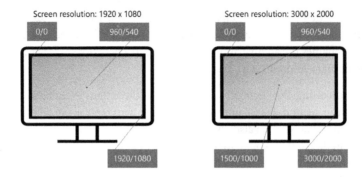

Figure 7.13 – Different screen resolutions and the same coordinates

As we can see in the preceding diagram, the coordinates **X=960** and **Y=540** are exactly in the middle of the screen on the left side. But when the screen resolution changes, these coordinates are in the top left area rather than in the middle. This means that when using actions that work with mouse coordinates, you must always make sure that the screen resolution is also the same as the one used to create the flow. With the **Get screen resolution** and **Set screen resolution** actions, these settings can be checked and changed if necessary.

Let's take a look at the options in detail.

Workstation

This action group comprises all actions that have to do with the workstation in the narrower sense. These include the following:

- **Workstation: Lock Workstation/Log off User/Shutdown computer**: It is important to distinguish the different actions. **Lock** corresponds to the same locking function in Windows, also available via the *Windows* key + *L* shortcut. The flow will continue to run afterward in the background. Logging off the current user instead will terminate the flow from that point on, as will the **Shutdown computer** action. The last one also has a parameter to define whether to just shut down, restart, hibernate, or set the workstation to sleep.

- **Print: Print document/Get default printer/Set default printer**: These work with a text field for the document name, getting or setting the printer.

- **Screen: Get screen resolution/Set screen resolution/Control screen saver**: These actions need a connected console session to work with. This session is available when a user is logged in to the workstation and PAD is executed by this user. That means that a user is logged in and has started the PAD console. In unattended scenarios, these actions don't have any effect. Setting the screen resolution can help to run scenarios in a standardized way, which works with screen coordinates, for example, mouse movements (see next section).

- **Miscellaneous other actions**: **Show desktop/Play sound/Empty recycle bin/Take screenshot**: It is worth mentioning here that **Show desktop** can both show the desktop and restore all windows. Furthermore, **Take screenshot** is useful for collecting and documenting information about a system for the entire desktop or foreground window.

Before we look at another example, let's first look at the actions for the keyboard and mouse.

Mouse, keyboard, and clipboard

The actions in this group complete the possibilities from this chapter in terms of simulating user sessions and provide all the actions that a real user can perform, starting from specific mouse sequences to sending shortcuts and predefined key combinations. Here are the details:

- **Block Input**: This action has a switch to block or unblock any keystroke or mouse click made by the user. This is useful so that a flow is not disturbed by additional keystrokes or mouse clicks from the user during execution. But be careful. On the one hand, please do not forget to insert an `Unblock` action at the end of the flow in any case. Also, please do not create a breakpoint in between because this will stop the flow, and there is no chance to continue the flow because of blocked input options.

- **Get mouse position**: This stores the coordinates of the mouse pointer in two variables (the x position and y position) relative to the screen or a foreground window.

- **Move mouse / to image / to text on screen (OCR)**: Can be used to move the mouse pointer to a specific position on the screen. You could also place the mouse pointer in the desired position and press *Ctrl* + Shift. This transfers the current coordinates into the action parameters. There are also multiple motion behaviors from **instant** to **with animation (high speed)**. When using the **Move mouse to image** action, it is possible to define or select an image (see *Chapter 5* on images) to move the mouse to and even to send a click after the move. PAD can also detect text on the screen by using **optical character recognition** (**OCR**) technology. The **Move mouse to text on screen (OCR)** allows you to select an OCR engine, text that should be found, and some additional options. For example, if you have identified an area of the screen where you want to perform a mouse click, which may be labeled with text, this action can be used successfully to move the mouse to that area and trigger the click.

- **Send mouse click**: This action can send a wide range of mouse events starting from standard left- and right-click, double-click, and middle-click. It is also possible to just send the button down event and the button up event at another time. This action also includes a switch that allows moving the mouse pointer to specific coordinates, which is essentially the same as the **Move mouse** action.

- **Wait for mouse**: When an application receives a mouse click, the mouse cursor might change to a **wait** cursor or an **hourglass** cursor. This action can be used to wait for the cursor to become/not become a regular cursor again.

With these mouse control actions, it is possible to simulate hardware that may not exist in front of the workstation, for example, if there is no middle button on the mouse. We can also implement and simulate *click-and-drag* or *drag-and-drop* operations. However, most applications can be operated not only by the mouse but usually also have a whole set of shortcuts that are entered directly from the keyboard and immediately trigger an action. To enable PAD to also use this very productive way of operating, there are the following keyboard actions:

- **Send keys**: If we want to type some text to the current location of the cursor in the foreground window/window instance/title or class of a window or a UI element, this would be the action to use. The following screenshot shows the parameters for this action:

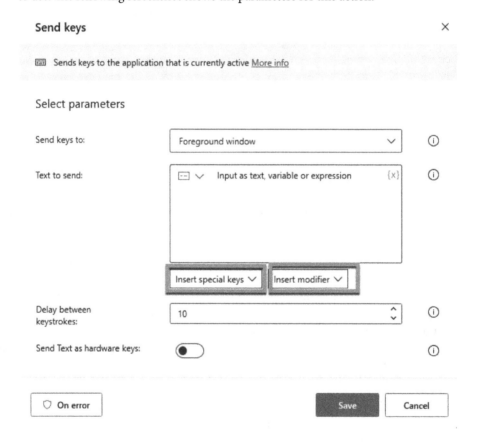

Figure 7.14 – The Send keys parameter dialog

The **Insert special keys** drop-down menu contains a list of keys that don't have a visual representation, such as *Enter* or *backspace*, but also comma, period, and space. In addition, we can select the arrow keys, function keys, and every key on the numeric pad. The **Insert modifier** drop-down menu is needed to create a shortcut command, which typically includes the *Ctrl*, *Alt,* or Shift buttons in combination with a letter. We all know the universal *Ctrl + X* (cut), *Ctrl + C* (copy), and *Ctrl + V* (paste) shortcuts. To send a shortcut to an application, for example, *Ctrl + C*, we would need to enter {Control}({C}) with curly brackets for C.

- **Press/Release key**: This action presses and holds the *Ctrl*, *Alt*, Shift and/or the *Windows* key or releases the keystroke. This means that between the press and the release, any other action or keystroke can be used.

- **Set key state**: This targets the **Caps Lock**, **Num Lock**, or **Scroll Lock** keys and sets the state to on or off.

- **Get keyboard identifier**: This receives the current keyboard identifier from the workstation's registry.

- **Wait for shortcut key**: This action allows us to pause the flow and wait for user input in terms of a shortcut. It is possible to define multiple shortcuts and use this input to enter different branches of the flow.

There is also an action group for the clipboard, which contains actions to get or set the clipboard text and to clear the content of the clipboard.

Now let's take a look into one example of how to use these actions.

Creating a PowerPoint system report by using mouse and keyboard actions

The following example uses some of the actions that we just learned about. We want to create a flow that collects some information about the current system, stores this in a PowerPoint slide deck, and saves that report on OneDrive. This would require a series of mouse clicks and key sends for this operation.

Please notice that we are not using UI elements for this example, which would also be a very good option. Instead, we want to simulate the user session with the previously defined actions. So, let's take a look at what the flow looks like:

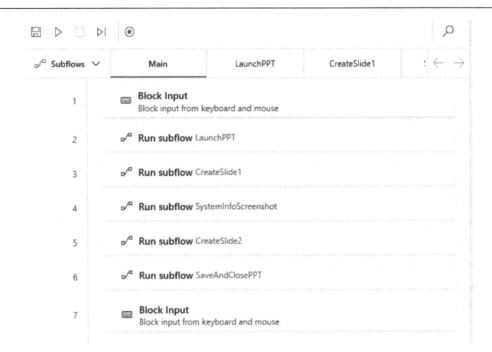

Figure 7.15 – A flow to create a PowerPoint system report

As we can see here, the flow is again structured into some subflows for easier readability and maintainability. The first subflow opens PowerPoint by using the Windows button, and the second one creates the first introduction slide with the current date. The third subflow launches the Windows **System Information** pane, takes a screenshot, and puts this into the clipboard. In the fourth flow, this screenshot is inserted into a second slide, and after some formatting, we save and close the slide deck in the default OneDrive folder. All subflows are embedded in a **Block Input** action, where the first turns on blocking and the second unblocks it.

Let's see the details of the first subflow:

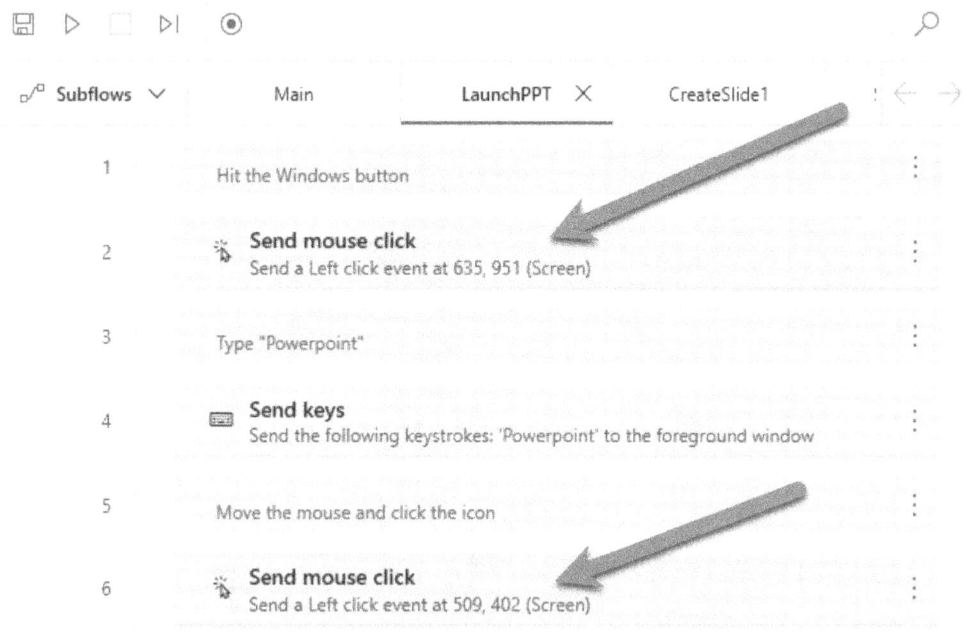

Figure 7.16 – Launch PowerPoint with the Send mouse click action

In the first action, we trigger a click on the Windows button to activate the **Start** menu. The coordinates shown here are for my desktop. They are probably different on another desktop.

> **Different desktops for flow creation and execution**
>
> We will later see that it is possible to use one computer for the flow creation and potentially another one (or multiple) for flow execution. When the **Send mouse click** action is used in a UI flow, it is recommended to make sure that the computer that is used for creation has the same screen resolution as the one which is used for execution (standardized environment). Otherwise, the coordinates for the **Send mouse click** action might not work. Alternatively, we could also use the actions for the screen resolution to establish standardization for the executing desktops

In the second action, we simply send the `Powerpoint` text to the **Start** menu using the keyboard input, which then causes the search for the corresponding program. In the last step, we again use a mouse click to launch the first program in the search list. Here, the comments about coordinates also apply. This will start PowerPoint, and we can move on to the next subflow:

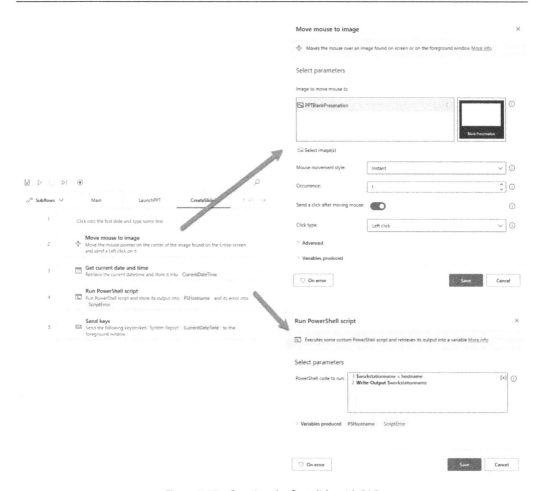

Figure 7.17 – Creating the first slide with PAD

In the first action, we click on the image for the standard layout in PowerPoint. The **Move mouse to image** action allows me to define an image that I just created within that action. This now results in PowerPoint displaying a presentation with the default layout and the first starting slide.

The next two steps determine the current date and the name of the workstation. For the latter, the scripting action for PowerShell was used. In the last step for this subflow, the heading consisting of the **System Report** text and the current date are inserted into the slide. We will use the workstation name later when saving the file. And that already brings us to the next subflow:

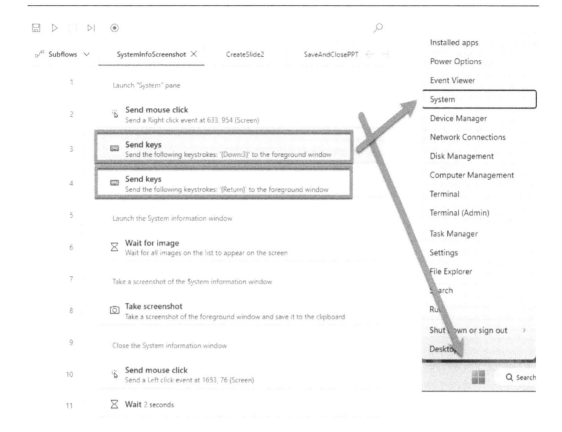

Figure 7.18 – Create a screenshot for System information with PAD

This subflow again uses the **Send mouse click** action to access the Windows **Start** menu, but this time, we use the right-click event to launch the special menu for administrative programs. In that menu, we use the **Send keys** action to trigger the down arrow key three times, followed by the *Return* key to launch the **System Info** pane. Next, we use the **Wait for image** action from *Chapter 4* and then take the screenshot of the foreground window into the clipboard before we close the **System Info** window again. Now, we have enough information to continue with the next slide in our presentation in the next subflow:

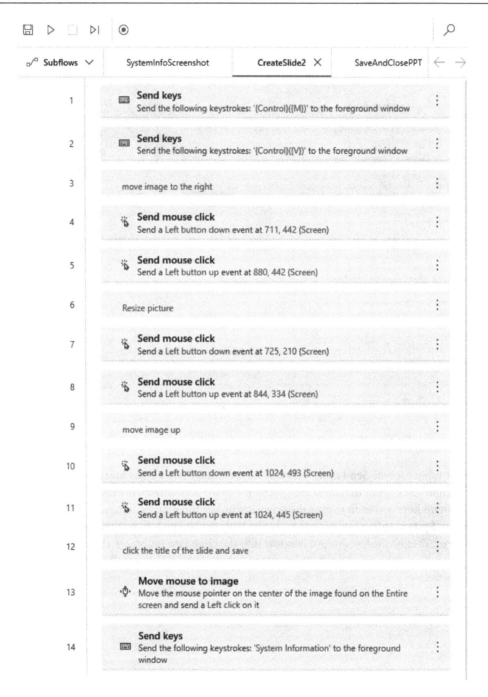

Figure 7.19 – Paste and resize a screenshot in PowerPoint

The first **Send keys** action uses a special keystroke. Using the keyboard shortcut *Ctrl + M*, we can insert a new slide into the current slide deck. To trigger this keystroke in PAD, we need to specify `{Control}({M})` with curly brackets for M. The same goes for the standard command *Ctrl + V* for pasting content from the clipboard into the current location.

In lines 4 and 5, we make use of the **Left button down** and **Left button up** mouse events to move the screenshot a bit to the right side of the slide. The same technique is then used to resize the picture (lines 7 and 8). We are making the image a bit smaller to fit better into the slide. To move the picture up, again, this technique is used in lines 10 and 11. Lines 13 and 14 fill in the heading information for this slide, which marks the end of the creation. The last subflow takes care of saving and closing PowerPoint:

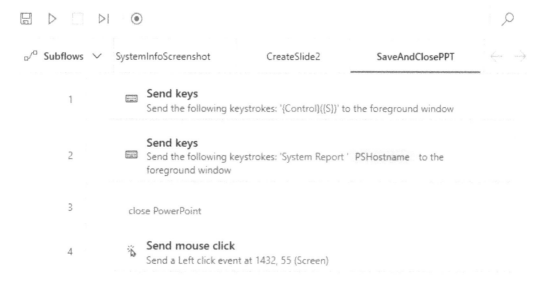

Figure 7.20 – Saving and closing PowerPoint with PAD

Again, the **Send keys** action is used to issue a keyboard shortcut to save the slide deck. This causes PowerPoint to display the **Save** dialog. Here, we just send the keystrokes for **System Report** and the workstation name that we extracted in our PowerShell script earlier. The last **Send mouse click** action just closes the program.

Final remark

The flow shown here was created on a desktop with a screen resolution of 1,920x1,200 with a scaling of 100% (i.e., no enlargement of the font or icons). If the flow were run on another desktop with a different resolution, then this would probably not work. In this respect, the flow shown here is a demonstration of what is possible. In practice, other mechanisms would certainly be used to create such a report.

Summary

In this chapter, we learned about a large number of actions that allow us to manage workstations. Furthermore, more information is available through scripting integration, and it is possible to execute more operations than would be possible through PAD alone. Also, the use of a mouse and keyboard extends the reach of PAD in the use of any program. Building on this, in the next chapter, we will look at special applications, including SAP and mainframe automation.

Further reading

- Command-line switches for Office programs: https://support.microsoft.com/en-us/office/command-line-switches-for-microsoft-office-products-079164cd-4ef5-4178-b235-441737deb3a6

- Command-line reference: https://learn.microsoft.com/en-us/windows-server/administration/windows-commands/windows-commands

- PAD actions reference: https://learn.microsoft.com/en-us/power-automate/desktop-flows/actions-reference

8

Automating Standard Business Applications

So far, we have encountered numerous examples in which we have integrated and automated various applications with PAD. This chapter serves to deepen some of these things and also to connect new widespread business application areas using PAD. This chapter will cover the following topics:

- Automating Microsoft 365 (also known as Office) applications

- Integrating and automating SAP

- Connecting and working with mainframe applications

- Automating browser-based business applications

By the end of this chapter, we will know how to automate two very common business application scenarios so that you can connect other applications. Furthermore, you will have a deeper understanding of how to automate web-based business applications.

Technical requirements

The sections in this chapter cover how to automate different software products, each of which also has different prerequisites. The details of these prerequisites are listed at the beginning of the respective sections.

The last section of this chapter provides an example of how to automate a web-based business application such as Dynamics 365. You will need to request a trial instance of Dynamics 365 for Sales. This trial is free of charge and quite sufficient for this exercise (you can request the trial here: `https://dynamics.microsoft.com/en-us/dynamics-365-free-trial/`). However, since this is a business solution, you will also require a Microsoft business account to request a trial version.

Automating Microsoft Office

In the previous chapters of this book, we used almost all Office products in our examples. We used Excel and Outlook in the introductory example in *Chapter 1*, while in the last two chapters, we saw how Microsoft Access and PowerPoint can be used as part of UI automation. While the latter concept can be used in conjunction with mouse and keyboard actions to include any application, there are dedicated PAD actions for Outlook and Excel that can be used within flows. We'll take a closer look at these action groups in this section.

Automating Outlook with actions

Being able to integrate Outlook into PAD can be very helpful in many situations. As already seen in our example in *Chapter 1*, it is possible to receive many emails or attachments of the same nature, move them to a specific folder, save them, and even reply if necessary. Another scenario could be that you retrieve messages from a support address, for example, and use that information to trigger further internal processing, such as entering certain data into other systems. The example in this chapter will show you another use case. Here, new users must be regularly provided with information so that they can log on to an SAP system. Here, too, Outlook can be integrated directly into a flow and send this notification directly. The following actions are available in the corresponding *Outlook* actions group to help us with these tasks:

- **Launch/Close Outlook**: This action starts or ends an Outlook instance. A corresponding variable is created in the **launch** action, which can be used in the **close** action later to terminate the program.

- **Retrieve email messages**: This action can accept an email account and allows you to define filter parameters for the emails to be received. In addition, settings can be made for any attachments.

- **Send a message/Respond to a message**: These actions create a new message with all the required fields, such as email body, subject, and attachments, and send it out directly. The **Respond to** action event has an additional parameter for the response action (reply, reply all, forward).

- **Process email messages**: This action allows you to move messages to another folder or delete messages that have been retrieved by the **Retrieve email messages** action.

- **Save messages**: This action stores email messages locally in a folder. It is also possible to define the format for the messages (text only, Outlook template, and so on).

To see these Outlook actions in action, please refer to the example in the next section.

Automating Excel actions

Just like Outlook, Excel has been used in some examples up to this point. Excel still enjoys great popularity among many users who use the program for list management and various calculations up to the database level. Perhaps this is also the reason why Excel has the most actions by far. If we expand the corresponding action group (**Excel**), we will find 11 actions in the main branch and 18 actions in the **Advanced** sub-branch. The list of functionalities is very extensive, from working with worksheets (add, delete, or rename) to cell editing (select, select ranges, delete, paste, copy, insert, or delete entire columns or rows) to special functions such as search and replace and even running Excel macros. So, if it is our task to create an even more complex Excel sheet over and over again and feed it with data, no wishes remain unfulfilled here. In such a case, it is also very practical to have corresponding actions for Excel, since this would hardly be manageable via the definition of UI elements.

Later in this chapter, we will see another example of how Excel can be used to control decentralized user management. This is the topic of the next section.

Working with SAP

SAP is both the company and the enterprise management software. At least by name, this software product should be known to everyone, since it is very widespread around the world, especially since the company represents the world's fourth largest company in the field of software and programming. For a very long time, **SAP ERP** was the standard software in the **enterprise resource planning** (ERP) area, which, in turn, was based on **SAP R/3**, which was first introduced in 1992. The latest version of SAP ERP is version 6.0, which was introduced back in 2006 and has been maintained with service packs and small enhancements since then. It is now considered legacy technology and the new version is **SAP S/4HANA** (**High-performance ANalytical Appliance**).

A while ago, however, SAP announced that SAP ERP will continue to be supported until 2027 or 2030 (with extended support; see the *Further reading* section for details), and certainly for good reason. This migration and switch to the new version, S/4HANA, is associated with high costs for many customers. Apart from the software, the hardware usually also has to be replaced, which can be very costly for ERP software if cloud use is not planned at the same time – all of which are reasons why we still find many SAP R/3 installations to this day. Automation with PAD can be very useful because companies think very carefully about whether and how much they still want to invest in an outdated software system. So, why not use a lightweight integration solution and put the money that's been saved into the project for the future ERP system? So, let's take a look at how we can use PAD to connect to the SAP interface, called **SAPGUI**, which needs to be installed locally for the following example and which is the standard client software for SAP ERP systems. This software is contained in the trial version of an SAP test system for development purposes, which was built up and used for this example.

Hint

Scripting must be installed and enabled for PAD to capture and target elements in this program. The following screenshot shows how to allow scripting within SAPGUI:

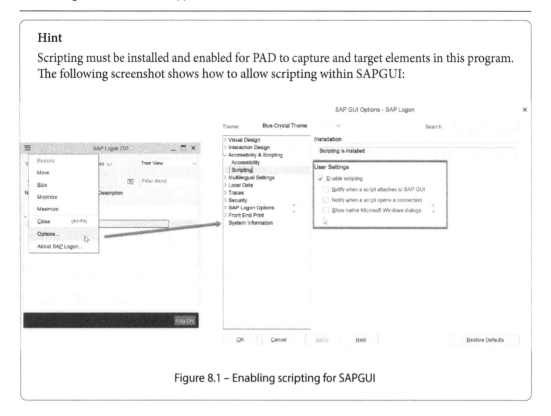

Figure 8.1 – Enabling scripting for SAPGUI

Distributed user management with PAD

Let's imagine that we are responsible for setting up users who will start in our small company next month. Besides a login name and a mailbox, of course, an SAP login is also needed. The HR department has already created accounts in Active Directory and collected the users for them in an Excel spreadsheet. Our task is to create the users listed there in SAP and to send the login information by email. This also includes creating an initial password. So, we need a flow that does the following for us:

- Read all users from the Excel spreadsheet and extract data

- Launch SAPGUI and call the corresponding transaction to create new users (SU01)

- Iterate through the Excel data, generate a new password, store this in the corresponding row of the table, and use the data to create a user in SAP

- Send an email to the user with the initial password

- Store the Excel worksheet and close Excel and SAPGUI

Since we don't have separate actions for automating SAPGUI, we have to use the techniques we already know to control this application remotely. However, since this application can also be operated excellently via keyboard input, we want to avoid creating UI elements and images as much as possible. Using keyboard input as the option to navigate through the application can sometimes be an advantage because this way, the user would not rely on screen resolutions and the need to capture all UI elements upfront.

Also, this time, we want to apply our proven practice of dividing the overall flow into smaller sub-flows. The complete flow will look like this:

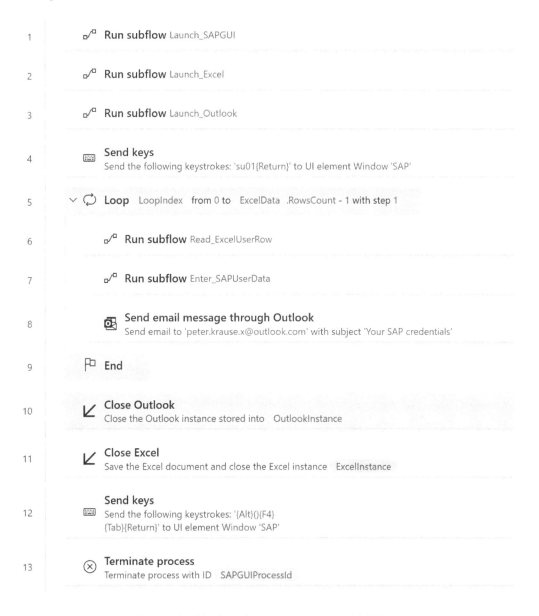

1   **Run subflow** Launch_SAPGUI

2   **Run subflow** Launch_Excel

3   **Run subflow** Launch_Outlook

4   **Send keys**
     Send the following keystrokes: 'su01{Return}' to UI element Window 'SAP'

5   **Loop** LoopIndex from 0 to ExcelData .RowsCount - 1 with step 1

6   **Run subflow** Read_ExcelUserRow

7   **Run subflow** Enter_SAPUserData

8   **Send email message through Outlook**
        Send email to 'peter.krause.x@outlook.com' with subject 'Your SAP credentials'

9   **End**

10   **Close Outlook**
     Close the Outlook instance stored into OutlookInstance

11   **Close Excel**
     Save the Excel document and close the Excel instance ExcelInstance

12   **Send keys**
     Send the following keystrokes: '{Alt}(){F4}
     {Tab}{Return}' to UI element Window 'SAP'

13   **Terminate process**
     Terminate process with ID SAPGUIProcessId

Figure 8.2 – Distributed user management with PAD

Now, let's examine the different sub-flows that are the most important and see what these are doing.

The Launch_SAPGUI sub-flow

As the name suggests, this area starts SAPGUI, which must be installed on your computer. The program is called directly in the first action. The program consists of two windows – one for selecting the system to log in to and another for logging on and running programs. To be able to address these windows in the further process directly, these are defined in the form of UI elements. To do this, it is a good idea to start the program once manually and define the two windows as UI elements via PAD.

To start the SAPGUI program, we could implement a flow like this:

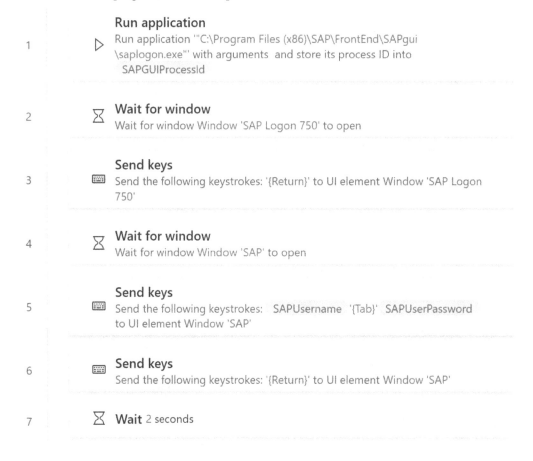

1   **Run application**
Run application "'C:\Program Files (x86)\SAP\FrontEnd\SAPgui
\saplogon.exe'" with arguments and store its process ID into
SAPGUIProcessId

2   **Wait for window**
Wait for window Window 'SAP Logon 750' to open

3   **Send keys**
Send the following keystrokes: '{Return}' to UI element Window 'SAP Logon
750'

4   **Wait for window**
Wait for window Window 'SAP' to open

5   **Send keys**
Send the following keystrokes: SAPUsername '{Tab}' SAPUserPassword
to UI element Window 'SAP'

6   **Send keys**
Send the following keystrokes: '{Return}' to UI element Window 'SAP'

7   **Wait** 2 seconds

Figure 8.3 – Starting SAPGUI with PAD

First, we must run the main `saplogon.exe` executable through the **Run application** action (line 1). This executable is available after the standard installation of the SAP client software. This is essentially the same as double-clicking on the corresponding shortcut on the Windows desktop. In line 2, we wait for the window to appear; after that, we will see the **SAP Logon** screen, where we can select a system to log on with:

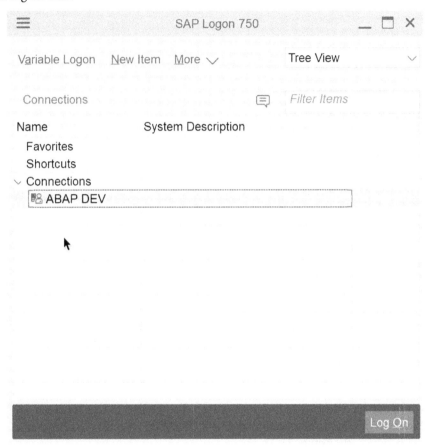

Figure 8.4 – SAP system selection

As we can see, I only have one system registered. That's why I just need to press *Enter* here by sending a keystroke (line 3). If you have multiple systems registered and want to log on to a specific system, you will need to make sure that the right entry is selected first.

In line 4, we wait for the main SAP window to appear, which looks like this:

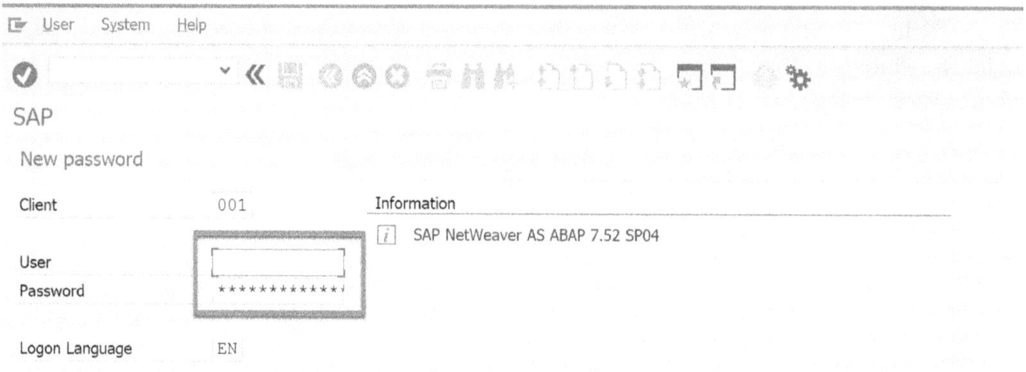

Figure 8.5 – SAP logon screen

As mentioned earlier, this software has also been designed to be used just using keystrokes. We can see where the cursor is located by the light red corners of the UI elements – in this case, the text box for the username. Again, this is the default behavior of that software and we could just make use of this. However, if we need to log on to a different client (see the **Client** field in the preceding screenshot), we would need to make sure that these settings are made.

In lines 5 and 6 in our flow, we enter the username and password. To jump from one field to the next, we can just use the *Tab* key. These values are stored in two input variables and are marked as sensitive. The last step is to press *Enter* and wait for the login procedure to finish (lines 6 and 7). At this stage, we have successfully logged on to the system and can start working with programs in SAP.

The "Launch_Excel" and "Launch_Outlook" sub-flows

The next steps in our main flow are to look up the users that we need to register in our SAP system and then send them their user information via email. Therefore, we must start Excel and Outlook with the corresponding actions that PAD provides for this.

It's worth mentioning that we start Excel with a given document that contains the list of users that we need to work on. We assume that this list is always in the same format and that we know where to look for specific entries. What we do in the launch routine is select and read the whole table of users (see the following screenshot):

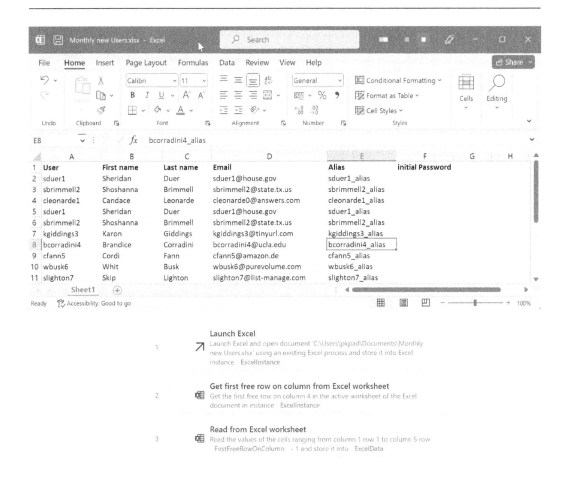

Figure 8.6 – Excel user list and PAD launch routine

We can do this by using the **Get first free row on column** action. This produces the
`FirstFreeRowOnColumn` variable, which we use in the next action, **Read from Excel worksheet**.
At the end of this sub-flow, we have all the data required for the processing stored in the `ExcelData`
variable. Because we want to store a generated password in the corresponding row for a user, we do
not close Excel at this point. From here, we can start processing each user.

User data processing

Now that we've launched SAPGUI alongside Outlook and Excel, we are ready to process the data for
each of the users. The first thing we have to do is call a specific transaction in SAP called **SU01** – this
is exactly what we do in line 4 of our main flow.

SAP and transaction codes

When working with SAP ERP, we often talk about so-called reports and transactions, which are used to execute business requirements and processes. A transaction code (for example, SU01) starts a program for entering parameters and very often starts a report afterward. The transaction used here, for example, allows a username to be entered, which can either be searched for or created. Therefore, the process shown here represents a very typical routine for working with SAP.

The following screenshot shows the action parameter dialog for this step:

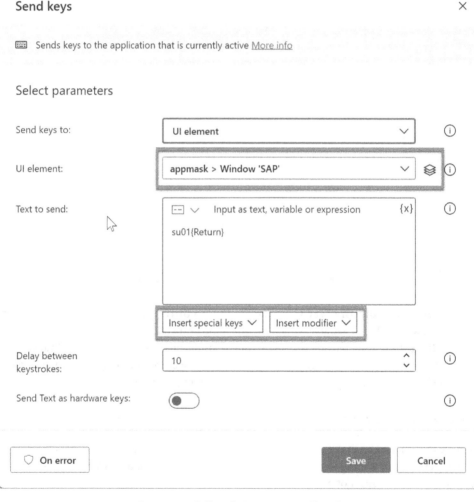

Figure 8.7 – Calling SAP transaction "SU01"

The reason we are looking at this step a little more closely here is that we will use this basic concept throughout this example:

- Because we can also select a UI window that we have previously defined with this action, we do not have to pay attention to whether this window is also in the foreground. This is especially useful when working with several programs in a flow and it is not always possible to know which window is currently in the foreground. By specifying the UI element, the keystrokes are sent to the specified window – in our case, SAPGUI.

- We can accomplish our task entirely through keyboard input, making us independent of additional UI elements such as buttons or list items. Thus, this flow also works independently of screen resolutions.

- The PAD action also allows us to perform special keystrokes such as *Shift + Tab* (jumps one field back in the interface) or *Ctrl + S* (shortcut to save, also in SAP).

- In addition to this, we can specify the time between keystrokes in milliseconds. This is useful when running SAPGUI on slower computers.

By executing this action, we will land on the screen for maintaining users in SAP. Next, we go through the loop (lines 5 to 9) in which we complete the following subtasks:

- Determine the user information to be processed (the `Read_ExcelUserRow` sub-flow). We read the table of users before.

- Using the information to enter it in SAPGUI (the `Enter_SAPUserData` sub-flow).

- Send an email to the user with initial login credentials.

For the loop itself, we are not iterating through the list of rows with the `For each` loop. Instead, we can use the index-based method because this makes it easier to write back data into the right row in Excel. Let's take a look at the details of the `Read_ExcelUserRow` sub-flow and see how this works:

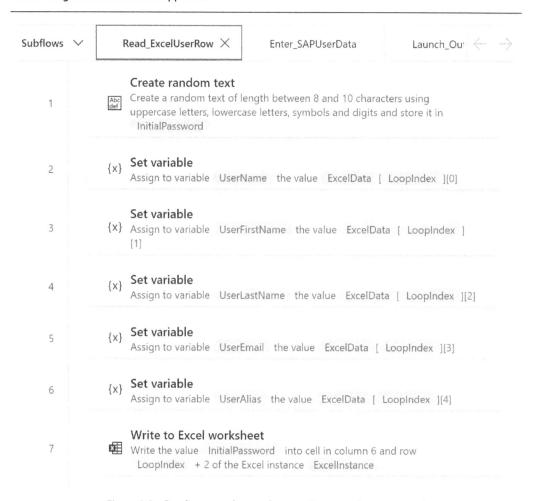

	Subflows ∨		Read_ExcelUserRow ✕		Enter_SAPUserData		Launch_Ou ← →

Create random text
1   Create a random text of length between 8 and 10 characters using uppercase letters, lowercase letters, symbols and digits and store it in InitialPassword

Set variable
2   {x}   Assign to variable UserName the value ExcelData [LoopIndex][0]

Set variable
3   {x}   Assign to variable UserFirstName the value ExcelData [LoopIndex] [1]

Set variable
4   {x}   Assign to variable UserLastName the value ExcelData [LoopIndex][2]

Set variable
5   {x}   Assign to variable UserEmail the value ExcelData [LoopIndex][3]

Set variable
6   {x}   Assign to variable UserAlias the value ExcelData [LoopIndex][4]

Write to Excel worksheet
7   Write the value InitialPassword into cell in column 6 and row LoopIndex + 2 of the Excel instance ExcelInstance

Figure 8.8 – Reading user data and generating a random password

Please note that we create a random password for the logon immediately in line 1. We use the **Create random text** action from the **Text** action group. This action was made for this purpose because possibilities such as length and complexity can be defined. In lines 2 to 6, we create flow variables to make it easier for us to work with this data later in the flow. Line 7 is responsible for saving the generated password in the corresponding line in the Excel worksheet. Here, we make use of the LoopIndex variable, which contains the number of the row we are processing right now.

In the next sub-flow, we are using SAPGUI again and operating the program to create a new user. As already mentioned, this is entirely possible using just the keyboard. The sequence we want to create can be represented by the following keyboard commands or inputs:

1. Enter a username, followed by *Tab*.

2. Enter a user alias, followed by *F8* (which is the function key for *create*).

3. Press *Tab* twice (jumps to the field for the last name).

4. Enter the user's last name, followed by *Tab*.

5. Enter the user's first name, followed by *Tab*,

6. And so on…

This sequence repeats for every new user and finally ends with *Ctrl* + *S* for **Save**, after which we find ourselves back on the screen to enter the next user:

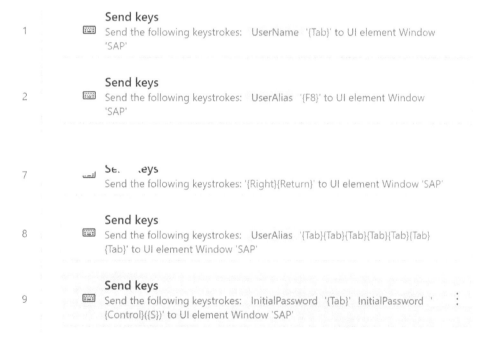

Figure 8.9 – The "Enter_SAPUserData" sub-flow

The last step in the loop is to send out the email to the user containing login information. For this, we must use the **Send email message through Outlook** action. Since we already opened Outlook at the beginning of the main flow, we could just use this action, which is configured like this:

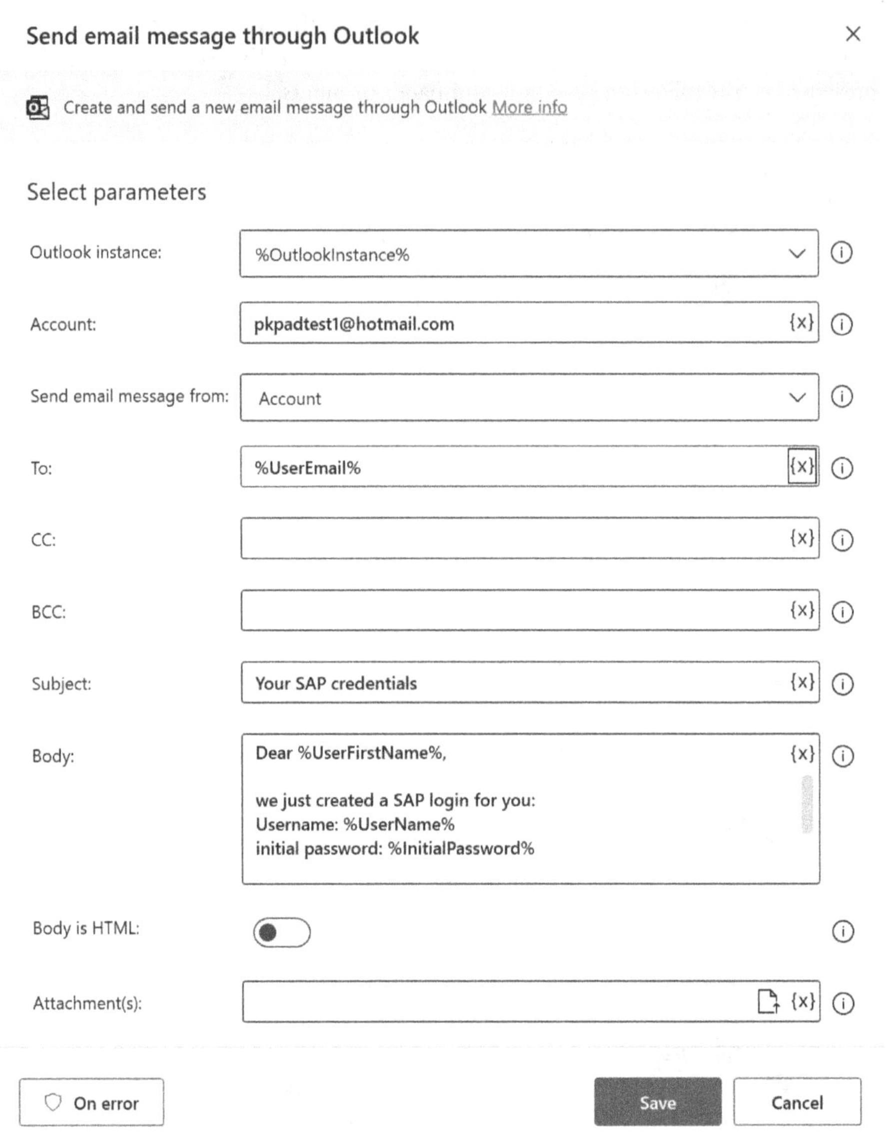

Send email message through Outlook   ✕

Create and send a new email message through Outlook More info

Select parameters

Outlook instance:	%OutlookInstance% ⌄ ⓘ
Account:	pkpadtest1@hotmail.com {x} ⓘ
Send email message from:	Account ⌄ ⓘ
To:	%UserEmail% {x} ⓘ
CC:	{x} ⓘ
BCC:	{x} ⓘ
Subject:	Your SAP credentials {x} ⓘ

Body:

Dear %UserFirstName%,   {x} ⓘ

we just created a SAP login for you:
Username: %UserName%
initial password: %InitialPassword%

Body is HTML:	⬤◯ ⓘ
Attachment(s):	⎘ {x} ⓘ

♡ On error   Save   Cancel

Figure 8.10 – The "Send email message through Outlook" action parameter

Here, we make use of the different variables that were created previously. They make the settings much more readable than using table variables with indexes.

At this point, the loop ends and starts over again for each of the records in that Excel workbook. After that, Outlook and Excel are closed via their corresponding actions. SAPGUI can be closed by a keyboard command that is common to all Windows programs: *Alt + F4*. After that, the program will want a confirmation to log off, which is also provided with another keystroke.

Additional thoughts and automation options

Of course, for every challenge, there are usually several possible solutions and the approach presented here is not the only one. It would also be possible, for example, to open Outlook only after processing the data in SAP and sending the emails to the users. This would probably also be the better variant for computers with little RAM. And as also mentioned, SAPGUI can be automated in other ways as well.

Microsoft has recently added extensive information about SAPGUI integration and automation to its documentation, which also underlines the importance of this possibility once again.

Here, we will talk about multiple implementation options:

- **Pro-code**: In this approach, VBScript is used to interact with the underlying SAPGUI Scripting API. Here, actions to be executed are recorded by SAPGUI in the form of VBScript commands, which can then be imported and used in PAD. This sounds simple at first. However, unless you happen to be an expert in this scripting language, changes in the process will be difficult to implement.

- **Low-code/no-code**: These two approaches use the recorder to capture specific fields such as buttons and text fields, which are tweaked afterward. This follows the concept of creating UI elements, which we learned about in *Chapters 5* and *6*.

This section covered a very popular business application where we learned how to use PAD to automate this sometimes very complex program. The next section looks at the beginning of all IT concerning business applications and shows how to automate and integrate mainframe applications, which are still greatly popular.

Integrating mainframe scenarios

The term **mainframe** refers to a family of computing machines that dates back to the 1950s. These are operated less by a graphical interface and more by a character-based user interface. They are characterized by very high reliability and very high throughput regarding input/output and arithmetic operations. Although the technology seems old-fashioned from today's point of view, mainframes are still widely used. In this area of classic IT, a distinction is made between batch processing and dialog programs. APIs do not exist in such environments. Information exchange is mostly file-based or via database access. Mainframe applications are most commonly used in the area of **enterprise resource planning** (**ERP**), financial accounting, warehousing, ordering, and more. Thus, many scenarios are conceivable in which integrating a mainframe application can be useful:

- A web store takes orders from customers, which have to be entered into the central warehousing system

- Suppliers use a web portal to confirm their deliveries, which, in turn, have to be entered into the central system

Especially when the entry has to be done via dialog programs, automation is a challenge, but one that we can tackle very well with PAD. There are several manufacturers of mainframes such as Unisys, Fujitsu Siemens, and, of course, IBM as the major one with their zSeries and AS/400 machines. One of the most commonly used operating systems is **Multiple Virtual Storage** (**MVS**), which was later renamed OS/390 and finally z/OS with 64-bit support for zSeries machines. Fortunately, for those of us who don't have a mainframe available or a connection to one of these, there is an emulation available for a slightly older version of MVS that runs very well on a Linux installation. In addition, on our PAD Windows machine, the terminal emulation software must be installed and network access to the emulation must exist. If everything has been configured, we can launch the terminal emulation and will see the welcome screen for MVS:

Figure 8.11 – Terminal emulation for accessing MVS

Software prerequisites

To control terminal applications, terminal emulation software is required, as well as access to a mainframe. PAD offers terminal emulation actions that work with Micro Focus Reflection or any other software that provides the HLLAPI implementation, such as RocketSoftware BlueZone, IBM Personal Communications, and Cybelesoft zScope. For this book, I have used a trial version of Micro Focus Reflection and also created a virtual machine with an emulation of a common mainframe system based on IBM technology. Please refer to the corresponding section for more information.

Now that we have all the prerequisites in place, we can log on to the system and explore its capabilities. To access programs, a valid user ID and password need to be provided. This version comes with pre-configured users and passwords that are provided in the documentation. The virtual design of mainframe scenarios is sufficiently complex. As a rule, users will find certain programs that fulfill a specific purpose, such as **update inventory** and **trigger supplier order**. These are organized in the form of libraries; end users can launch these programs directly via keyboard input and then start business processes with data and parameters. In our test environment, unfortunately, we do not have any programs that trigger business processes, but the principle is identical, so we can use an example flow to demonstrate this interaction pattern.

The list of actions available to us for terminal sessions shows that such applications are operated exclusively via the keyboard:

- **Open/close terminal session**: When **Open terminal session** is used, a variable is created containing the terminal session object. Here, it is possible to choose between different providers (*Micro Focus Reflection* or generic HLLAPI) and also the installation path and session to connect to. In my example, I created a session file containing the connection information and provided this to the parameters:

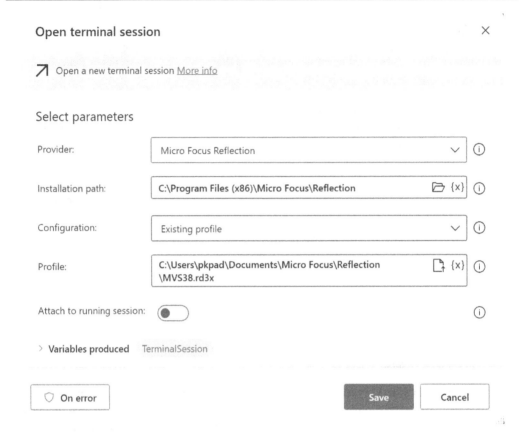

Figure 8.12 – The "Open terminal session" action parameters

- **Move cursor on the terminal session**: Sometimes, the cursor needs to be placed in a specific position to enter some data. This action can be used to place the cursor in the desired position by providing a value for the row and column number.

- **Get/set text from or to the terminal session**: These actions are used to enter or read data into the mainframe program – for example, the data of an order to be entered, including the order positions, could be entered with set text. Then, later, an automatically created order number could be read out with get text.

- **Send key to terminal session**: This action is used to perform a control within the terminal emulation – for example, by sending the Transmit command (equivalent to the *Enter* key) or other control commands to be able to switch between the different screens.

- **Wait for text on terminal session**: Similar to the UI elements, this action can be used to wait for a specific piece of text or screen to appear if there are latencies due to processing caused by entering parameters on the mainframe.

In the following example, we are using almost all of these actions to simulate a particular flow for a mainframe program. To do this, we simply call the system monitor and extract the value for a particular field and store it in a PAD variable. For demonstration purposes, the values for username and password, which are used to log on to the system, are hardcoded in the flow:

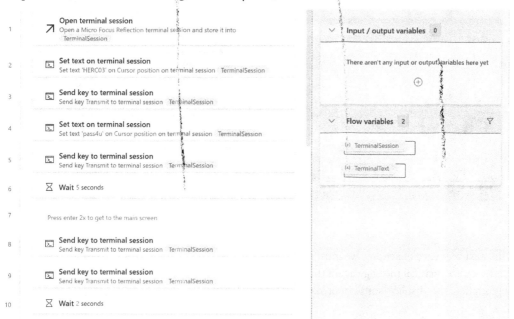

Figure 8.13 – PAD logging on to a terminal session

First, we need to open a terminal session (line 1). Here, the cursor will be placed at the position to enter a username. Next, we send the username to the terminal session and the <transmit> message (lines 2 and 3), followed by the password (lines 4 and 5). Following this, we must give the system some time to process before we send the <transmit> message again twice. We need to do this because two additional screens will be presented by the operating system (broadcast messages) before we land on the main screen:

```
Terminal  CUU0C0                                      Date  22.01.23
System    TK4-                                         Time  13:03:47
TSO User  HERC03

Option ===> 1

                  The MVS 3.8j Tur(n)key System
           TK4- Version 1.00 Update 08 -- MVS PUT 8505

                      TSO Applications

          1   RFE       "SPF like" productivity tool
          2   RPF       "SPF like" productivity tool
          3   IM        IMON/370 system monitor
          4   QUEUE     spool browser
          5   HELP      general TSO help
          6   UTILS     information on utilities and commands available
          7   TERMTEST  verify 3270 terminal capabilities

                      Enter X to Terminate

   PF3=Terminate

TT■                        »⇑                         0   5,14   A
```

Figure 8.14 – The MVS main screen

In the next sequence of actions, we must navigate to the **IMON/370** application by sending some text and the <transmit> message again (lines 12 and 13). This application is an interactive monitoring program that also displays some information about the system:

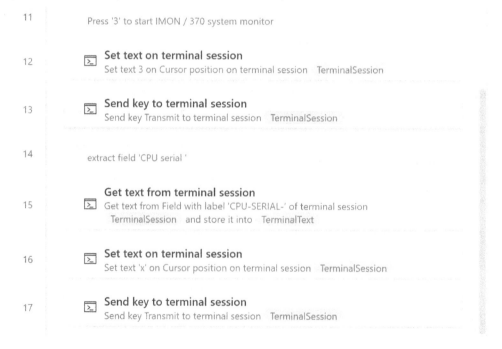

11   Press '3' to start IMON / 370 system monitor

12   Set text on terminal session
            Set text 3 on Cursor position on terminal session   TerminalSession

13   Send key to terminal session
            Send key Transmit to terminal session   TerminalSession

14   extract field 'CPU serial '

15   Get text from terminal session
            Get text from Field with label 'CPU-SERIAL-' of terminal session
                TerminalSession   and store it into   TerminalText

16   Set text on terminal session
            Set text 'x' on Cursor position on terminal session   TerminalSession

17   Send key to terminal session
            Send key Transmit to terminal session   TerminalSession

Figure 8.15 – Launching a program in MVS and extracting a value

In line 15, we extract the value for the **CPU-SERIAL** field into a PAD variable, before we quit the program again (lines 16 and 17).

To finish processing properly, we still need to log out of the terminal session:

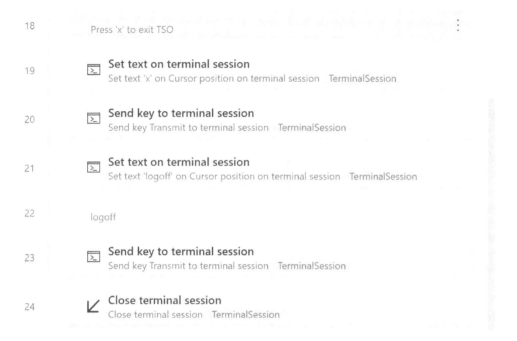

Figure 8.16 – Using PAD to log off from the terminal session

The first thing we must do is log off from the user session (lines 19 and 20) and then log off by sending the corresponding message, followed by `<transmit>`. Finally, we can close the terminal session (line 24).

In this section, we have seen that it is possible and not very complicated to automate mainframe applications. PAD provides all the necessary elements to call and run mainframe programs. In addition, results or calculations that have been created can be read out and further processed elsewhere, so a bi-directional interface is made possible.

In the next and final section, we will focus on an area in which most business processes take place today: browser-based applications.

Browser-based business applications

So far in this chapter, we have talked about business applications that are installed locally on a computer. Although there are many of these applications today, developing business applications for browsers has also become more popular and widespread. Of course, this also is related to the fact that many software products are developed and provided as **Software-as-a-Service** (**SaaS**) by major manufacturers. These include, for example, ServiceNow, Salesforce, SAP, Microsoft, and many more. As a rule, these software products also have an **application programming interface** (**API**) through which all functions and data are available. In the Microsoft ecosystem, we have Power Automate as part of Power Platform, where a constantly growing number of connectors can be used, for example, to establish a data connection to the aforementioned systems and to work with them.

But sometimes, the possibilities of API usage are not available and data must be entered via the interface. This is where PAD can help. For this use case, a pattern can be derived that applies to all browser-based business applications:

- There are no default actions for the applications in PAD; instead, the actions from the **Browser automation** action group can be used

- UI elements must be created for the flow to be mapped and to work with the interface

- For interaction (entering data, reading information, and so on) with the applications, the actions in the **Web data extraction** and **Web form filling** action subgroups can be used

Before we dive into an example, let's look at the different actions that we can use for browser automation.

Browser automation action group:

- **Launch new browser/close web browser**: PAD offers standard actions to open a new instance or connect to a running instance of Internet Explorer, Firefox, Chrome, or Microsoft Edge. These actions also allow you to clear the browser cache, clear cookies, and wait for a given web page to load with a given timeout. A variable called `Browser` is created to be referenced from other actions. The **Close web browser** action contains no surprises.

- **Create a new tab/go to web page**: This action expects a running browser instance that it can use to create the new tab with a specific URL or use to navigate. The **Go to web page** action can also use browser navigation for **Back** or **Forward** or even for **Reload**.

- **Click a link/download link**: The former action is probably one of the most used in these kinds of scenarios. To use it, a UI element must be selected. Then, there are different click types to choose from (a simple left click but also a right click, middle click, double-click, and separate types for left/right and down/up). To download a file to a specific folder, the latter action can be used.

- **Run JavaScript function on web page**: This action allows you to execute a self-written JavaScript function to be executed on the web page. Any JavaScript function can be injected here. One example could be to use this to scroll down on a web page to fill in or extract more data.

- **Hover mouse over element**: Sometimes, valuable information is already displayed when the mouse hovers over a certain element – that is exactly what this action is built for. A UI element has to be provided for this action to work.

- **Wait for web page content/If web page contains**: These two conditional actions also work with UI elements. To make sure that a certain element exists on a web page – for example, after a button was pressed and some processing has been triggered – these elements can be used to interrupt the flow's execution until a certain element appears (or disappears) on the web page. The second action can also check if a given piece of text is contained/not contained on the web page.

In addition to these general actions, there is also a subgroup for **Web form filling** and **Web data extraction**. To transfer data from PAD to the web application, we have actions to focus and populate a text field, set a check box, radio button, or a drop-down list value, and to press a button on a web page. These are all located in the **Web form filling** subgroup and also work with UI elements.

The **Web data extraction** subgroup contains actions to get more details about the web page itself or an element, as well as a generic **Extract data from web page** action that needs a bit more explanation. If we just drag this action into the flow designer, we won't be able to define any additional meaningful parameter. Instead, this action should be used in conjunction with the PAD recorder. We learned about it in *Chapter 2*, where we saw an example of how to capture a list of data from a web page.

This action is pretty sophisticated and allows us to extract single values, lists, rows, or even complete tables, depending on the structure of the web page we want to extract data from. We can describe a general procedure for extracting web data as follows:

1. Determine the URL of the web page to work on.

2. Start the PAD recorder and walk through the actions that should be performed. The recorder will also create corresponding UI elements that must be executed later.

3. Use the context menu for the web elements to insert data or capture data. A single click would just select the web element and capture this action, but by right-clicking on an element, there are more options, especially for data extraction.

4. Observe the list of recorded actions. The click sequence does not have to be perfect and if you accidentally clicked on an unwanted element, just delete it in the recorder.

5. Once you're finished, review the generated UI elements and give them meaningful names. It is much easier to identify the elements afterward in the corresponding actions.

6. Test your flow and modify the selectors. During the recording, the web elements are captured, but sometimes, they will need further tweaking. This can be done from out of the designer by editing the selector.

Now, it is time for a little example of how to use browser automation. For this, we are going to insert a new quote into Dynamics 365 Sales, one of the CRM modules of Dynamics 365. In this scenario, we want to use the Dynamics 365 Sales module to create a quote, read the generated quote number, and store it in a variable for further processing.

Like any other web-based business solution, Dynamics 365 has a few features to consider:

- A general setup and login are required for a user to work with the application. If you just started the trial, everything will be set up for you in the background. In the following example, the user who is logged on to the workstation is also logged on to the underlying Azure Active Directory instance, as well as Dynamics 365. This means that the browser window will not ask for a login separately. A login to the system may need to be considered independently – for example, via stored credentials.

  Optional: Do the authentication upfront before starting the PAD. This will ensure your automations work smoothly.

- A specific URL is used to access the system and that URL needs to be provided for the **Launch new Browser** action. For Dynamics 365, it is even possible to use dedicated URLs for specific business actions such as **create account** or **edit contact**. In this example, we will use the URL for creating a new quote directly.

With these prerequisites in place, the PAD flow shown in the following figure can be recorded right away. Here, we need to do the following:

1. Create a new flow and give it a meaningful name (for example, `Dynamics 365 Quote Creation`).

2. Once the designer shows up, press the recording button in the top button menu.

3. Perform the steps that are needed for the quote to be created. In our case, we need to enter a quote name, select a price list and a customer, save the record, and extract the generated quote identifier.

4. Beautify the generated UI elements and variables to make this a little bit more readable.

In the following figure, we can see the flow that could be generated out of this on the left-hand side and the improved one on the right-hand side:

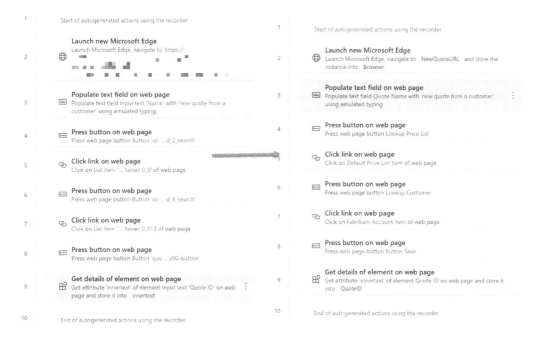

Figure 8.17 – Creating a new quote in Dynamics 365

In addition to renaming the UI elements and variables, there is one thing to take care of when working with this application. In line 3, we are populating the quote name with some text. For this action, we need to disable the option for populating the text using physical keystrokes and instead use emulated typing, which is typically required on complex web pages:

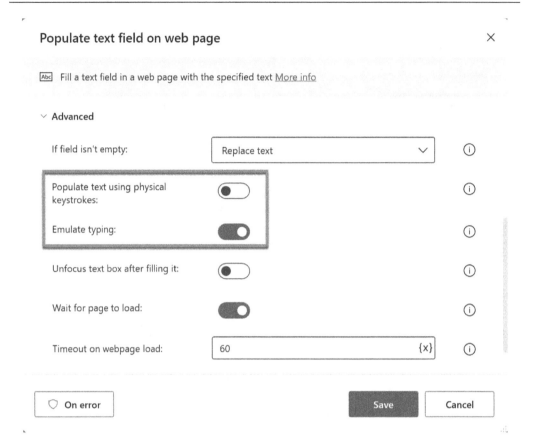

Figure 8.18 – Using emulated typing on complex web pages

For all other UI elements, the recorder does a very good job and the flow can be reliably executed repeatedly. Please note that the flow can be enriched afterward by further actions in the designer. This is the usual procedure to insert a **Wait for web page content** action, for example. As a result, a new quote will be created in Dynamics 365:

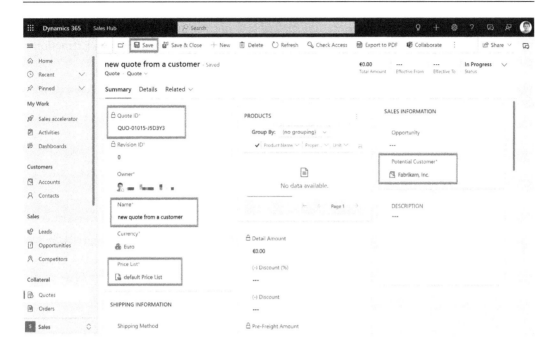

Figure 8.19 – A new quote created by PAD

Notes on this example

At this point, we could ask ourselves whether it makes sense to use this method to perform actions in Dynamics 365. There may be several reasons for this:

- There is currently no one else available to feed us this data via another interface, such as a cloud flow

- This flow is possibly only a sub-flow and other systems have to be served locally

In addition, it should be noted that due to space limitations, we have only mapped a portion of the flow here. Of course, a quotation also consists of quotation items, which would naturally have to be taken into account and considered accordingly. Information such as the name of the offer, the customer, and also the price list would be provided via variables in a real scenario.

A business application such as Dynamics 365 has a sufficiently complex HTML structure. The interface is mostly the result of a rendering process performed by an underlying framework. In Dynamics 365, for example, the React framework is used for this purpose. However, this also means that the selector for an HTML element can have a very complex structure. So, if you have any doubts, some manual work is required to find the right element. More about this will be provided in *Chapter 12*.

The ideas and concepts presented here are also applicable to normal and public websites. As a rule, simpler structures can be found here, which are also better recognized by the **Web Data Extraction** action – for example, HTML tables.

However, this example is already a good transition to our next chapter, in which we will take a closer look at the capabilities of Power Platform, which is also home to Dynamics 365.

Further reading

To learn more about the topics that were covered in this chapter, take a look at the following resources:

- Innovation Commitment for SAP S/4HANA, clarity and choice on SAP Business Suite 7: `https://support.sap.com/en/release-upgrade-maintenance/maintenance-information/maintenance-strategy/s4hana-business-suite7.html`

- *Introduction to SAP GUI–based RPA in Power Automate Desktop* (contains video) – Power Automate | Microsoft Learn available at `https://learn.microsoft.com/en-us/power-automate/guidance/rpa-sap-playbook/introduction`

- Mainframe computer: `https://en.wikipedia.org/wiki/Mainframe_computer`

- MVS: `https://en.wikipedia.org/wiki/MVS`

- MVS 3.8j Tur(n)key 4 – System: `https://wotho.ethz.ch/tk4-/`

- *Browser automation actions* reference – Power Automate | Microsoft Learn available at `https://learn.microsoft.com/en-us/power-automate/desktop-flows/actions-reference/webautomation`

- *Install Power Automate browser extensions* – Power Automate | Microsoft Learn available at `https://learn.microsoft.com/en-us/power-automate/desktop-flows/install-browser-extensions`

9

Leveraging Cloud Services and Power Platform

In this chapter, we will cover how to integrate PAD with Power Platform and learn how this technology can be used to bridge the gap between the cloud world and a locally executed UI flow. In addition, we will look at the actions to work with Microsoft SharePoint from a UI Flow.

Finally, we will work with the actions for **Amazon Web Services (AWS)** and Microsoft Azure to use the **Infrastructure-as-a-Service (IaaS)** offering for outsourced UI Flows. We will cover the following topics in this chapter:

- Power Platform and related services
- Managing lists, files, and folders with SharePoint
- Working with IaaS offerings (Azure/AWS)

Technical requirements

This chapter shows how to integrate and use a Desktop Flow with Microsoft's Power Platform. This scenario requires a Work or School account, which can be used to create a trial version of Dynamics 365. The example in this chapter uses the Dynamics 365 Sales module, which can be set up in just a few steps (a link for detailed information is contained in the *Further reading* section). Basic knowledge of calling and operating SaaS services is a prerequisite. Additionally, a license for Power Automate is required, which includes UI flows. This can also be a test license.

To showcase functionality with Microsoft Azure and AWS, a trial of both environments is required. The providers provide free offers for the limited use of resources, which are completely sufficient for our purposes.

In addition, basic knowledge of the providers and their offerings is required, as well as knowledge of how to create a simple standardized virtual machine.

Power Platform and related services

In *Chapter 1*, we learned that Power Automate Desktop is a part of Power Platform. With this direct relationship comes a unique advantage that can be used in conjunction with the other members of the Power Platform family. In the previous examples, we saw that a UI flow connects to another application based on specific use cases, or browser automation is used to retrieve information remotely. By using Power Platform, this scenario of outbound connections can now be extended to an inbound scenario – for example, a UI Flow can be called automatically from outside and do its work. In this respect, PAD can close the gap between cloud and local networks and thus ensure an interconnected process flow.

To gain a bit more understanding of this concept and how all this fits into the Microsoft ecosystem, I want to explain what all this means at a high level. The following figure shows the conceptual structure of the Microsoft cloud:

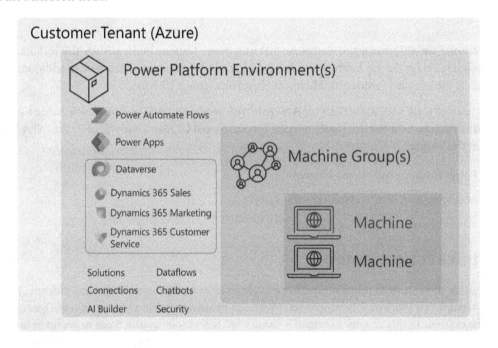

Figure 9.1 – Power Platform environment(s) structure and elements

If any customer wants to use a Microsoft cloud service, a customer tenant must be created. Here, we would find all the resources that are available in Azure but also the so-called **environments**. This term is related to Power Platform capabilities and can be considered a kind of container that serves a specific topic or purpose. There can be multiple environments in a tenant, such as for development, testing, and, of course, production, but also separate environments for HR workloads, internal service management, or any other topic.

Within each environment, we can find the different tools that are provided by Power Platform such as Power Apps, Power BI, Power Automate, and so on. Also, the access for users can be controlled for each environment separately. The preceding figure shows a selection of the services that can be available in an environment. This also depends on the specific license and subscriptions that have been purchased by the company. All artifacts of the environment are stored in a Dataverse repository. Power Automate is relevant to our topic in this chapter. Let's take a look at what is necessary to connect a cloud flow to a UI flow.

Machine and Machine Runtime

As we know, a UI flow is executed on a local machine, typically within the local network of a company. To be able to connect the Microsoft cloud and the environment to that local machine, Microsoft provides the **Power Automate Machine Runtime**. The following figure shows the relationship between the components:

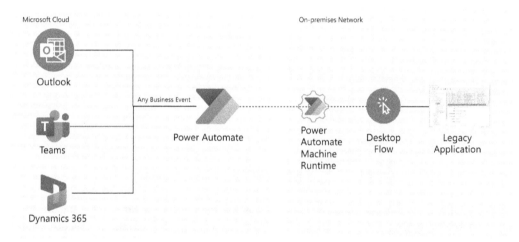

Figure 9.2 – PAD and Power Platform

This is installed automatically on Windows Professional and Enterprise, but not on Windows Home Edition. **Power Automate Machine Runtime** allows you to create direct connectivity between the cloud environment and the local network. To use this component, a couple of prerequisites are required:

- This type of connection is supported by Power Automate version 2.8.73.21119 or later
- The environment needs a specific solution called **MicrosoftFlowExtensionCore** of version 1.2.4.1 or greater
- A premium per-user plan with attended RPA is required, which is a special license that needs to be purchased
- The user that needs to register a new machine must have at least the **Environment Maker** or **Desktop Flow Machine Owner** user role in the environment

Installing the runtime is straightforward. The user with the appropriate privileges can select and register the local machine in the environment. The following screenshot shows a successfully registered machine:

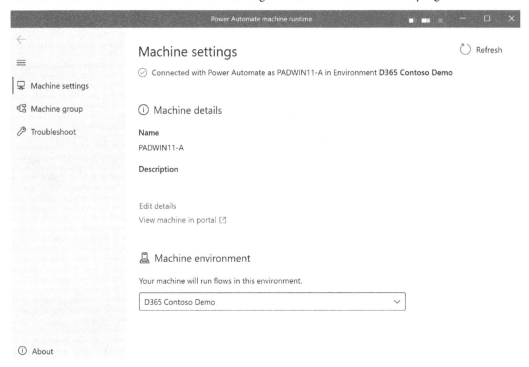

Figure 9.3 – Power Automate Machine runtime settings

If a high load is expected for one machine, it is also possible to define a machine group and add the machine there. Presuming that there are several machines available for execution, execution can be distributed among them. Creating a machine group is very easy. It can be accomplished via the Power Automate portal or within Power Automate Machine Runtime. Only a name and an optional description must be assigned. Then, the group is created and a random password is generated, which must be specified when adding further machines to the group. After that, the local machine can be added to the group, which will look like this:

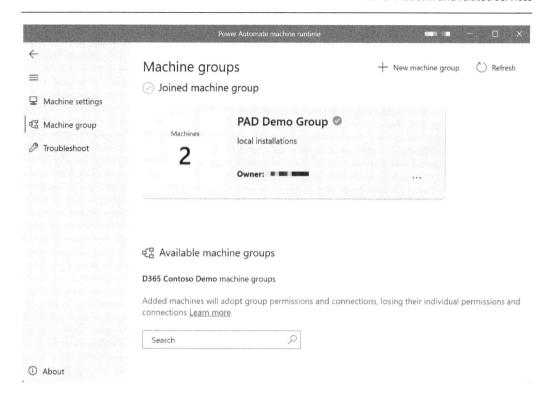

Figure 9.4 – A machine added to a machine group

There is of course also an online management console for these activities. To manage the configuration of Power Automate within Power Platform, do the following:

1. Open a browser and go to `https://make.powerautomate.com`.
2. Log in with valid credentials and make sure that the right environment is selected. There is an environment selector in the top-right corner of the portal.
3. Choose **My flows** in the left navigation to see a list of all types of Power Automate flows. There is a tab for Desktop flows that reveals all Desktop Flows since these are also stored here within the Dataverse repository.
4. Expand the **Monitor** node and click on the **Machines** entry. This will give you a list of the registered machines for this environment, as well as the machine groups:

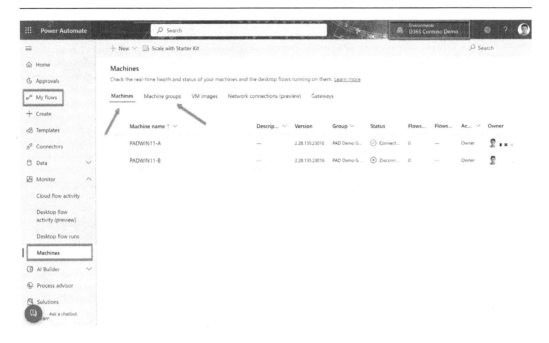

Figure 9.5 – The Power Automate Management portal

The portal also provides access to many other monitoring tools, and it is also the place to create new cloud flows, which can make use of Desktop flows.

Accessing a PAD flow from the cloud

To do this, we must switch to the **Create** entry in the left navigation of the Power Automate portal. Several ways to create a cloud flow will be presented. In the context of this example, let's consider the small scenario shown in the following figure:

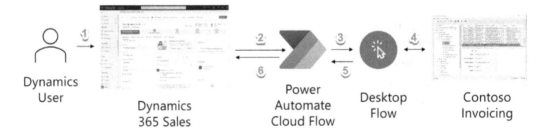

Figure 9.6 – Contoso Invoicing scenario

Let's take a look at this in more detail:

1. There are sales reps (Dynamics users) who maintain accounts and opportunities in Dynamics 365. A sales representative has created a new opportunity and the customer has accepted it. The opportunity is marked as **Won** and now needs to be transferred to a local application for billing.

2. A Power Automate Cloud flow is triggered on that event. The cloud flow collects additional information that is required for the invoice to be stored.

3. The cloud flow then calls the Desktop flow and passes over the required information.

4. The Desktop flow takes that information, launches the local application, and creates the invoice.

5. The Desktop flow also extracts the invoice ID and returns this to the cloud flow.

6. The cloud flow then creates an additional activity for the opportunity containing the invoice ID for the user to be informed.

For this scenario, we use the standard sales module of Dynamics 365. The local invoice management is mapped via Contoso Invoicing, which we got to know in a previous chapter. Please note that this is just a simplified sales process. The Dynamics 365 Sales module also includes data structures for quotes and many more that we are not focusing on right now.

The overall process is divided into two parts: the cloud flow and the Desktop flow.

The Desktop flow provides the following functionality, which should already be very familiar to us:

- Receive the incoming values and start the invoicing program
- Remotely control the process of entering the invoicing data
- Extract the generated invoice ID and return it to the cloud flow

Receiving and providing values is ensured via **Input / output variables**. To enter a new invoice into the Contoso Invoicing application, we need to provide an account name, a contact name, and an amount. Here, we come full circle to our chapter on variables. As shown in the following screenshot, three input variables and one output variable have been defined in the flow:

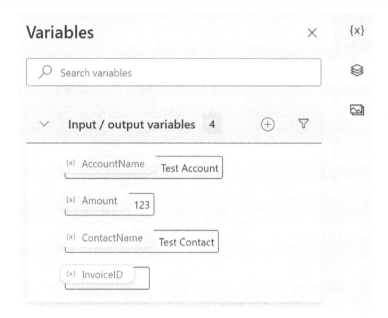

Figure 9.7 – Input/output variables for cloud communication

The InvoiceID variable is an output variable that will contain the generated invoice number from the application.

The rest of the UI flow was discussed in the previous chapters: working with variables and defining UI elements (*Chapter 5*), as well as using UI element actions (*Chapter 6*). A certain familiarity with the application to be automated is also required, of course.

Let's take a look into the cloud flow and what is required to call the Desktop flow from within. The cloud flow contains the following functionality:

- Trigger on opportunities that are closed as won
- Get details for the opportunity and related records and pass over all necessary values to the Desktop flow
- Wait for the Desktop flow to finish and receive the invoice number
- Create a task with the generated invoice number and attach it to the opportunity

The following figure shows what this cloud flow looks like:

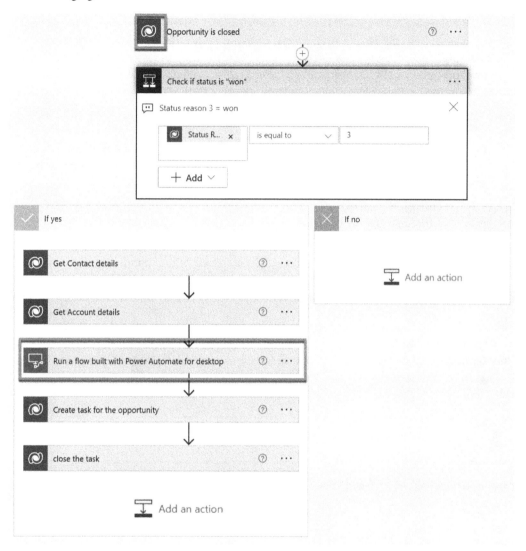

Figure 9.8   Power Automate cloud flow triggering a desktop flow

Without going into the details of Power Automate, here are some hints for creating such a flow:

1. A flow always needs a trigger. This can be a change in a record or a new email. A manual trigger is also possible. In any case, the triggering system must be determined using the corresponding connector. There are several hundred connectors for all kinds of systems and this number is constantly growing. In our example, we have used the connector to Dataverse, which is the basis for the Dynamics Sales module and is represented by the green symbol. This also means that, in principle, any other system can be used in Power Automate – for example, ServiceNow, Salesforce, or simply a database.

2. Power Automate allows you to control the flow through loops and conditions, shown as a gray box at the top of the figure. Here, we check whether the record has a certain status so that it can be branched accordingly.

3. In the Yes branch of the condition, additional information about the customer and the contact person is retrieved using the Dataverse connector, which is necessary for invoice generation. Afterward, the desktop flow is called (the blue box).

4. Once the desktop flow has finished, a task activity is created for the opportunity, which contains the generated invoice number, so that the sales representative also knows which invoice number was created for the sales transaction record.

When the cloud flow is initially created, appropriate connections are set up in the background for the connectors used. These contain credentials to the services behind them that are used with the connector. Accordingly, a connection must be created for desktop flows when they are used for the first time. To add this step, the *plus* sign between the two existing steps or the **+ New step** button in Power Automate Designer can be used, as shown in the following screenshot:

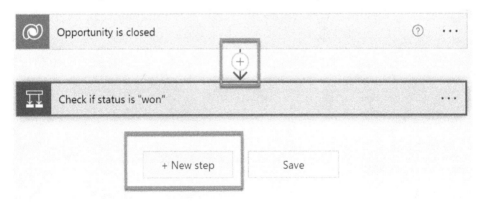

Figure 9.9 – Adding a new step to Power Automate

Next, the system to be included in the flow can be selected. In this example, the **Microsoft Dataverse** and **Desktop flows** connectors are being used:

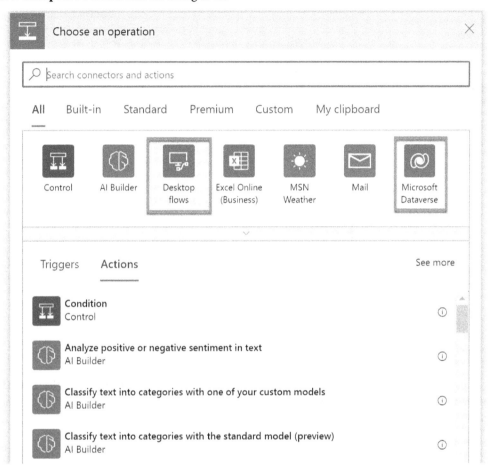

Figure 9.10 – The Desktop flows connector in Power Automate

After selecting the system, the available actions are displayed in the lower area of the window. In the case of Power Automate Desktop, the **Run a desktop flow** action is available. A window is then displayed in which the connection information must be entered. Here, you can also choose which machine/machine group is to be used for the execution:

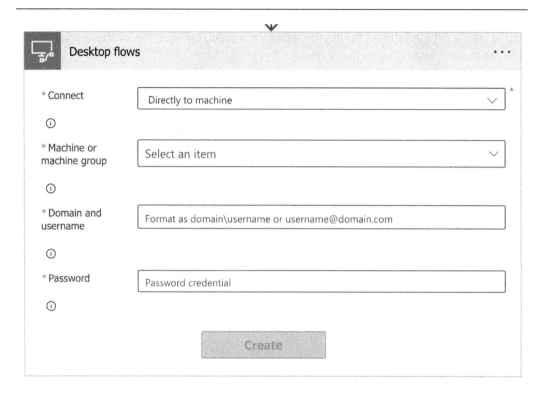

Figure 9.11 – Creating a connection for PAD

Afterward, the desktop flow, which is to be executed within the scope of this processing, can be selected. The so-called **Run Mode** is defined here (see the next section), as well as all input variables that the Desktop Flow would like to have for execution (see the following figure):

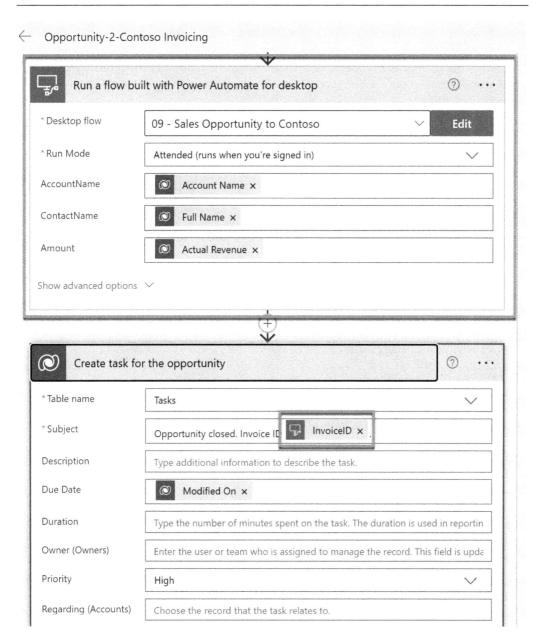

Figure 9.12 – Configured Desktop Flow action and parameters

Dynamic values from the entire cloud flow can then be used. In this case, the account name, the contact's name, and the amount from the respective dataset are passed. The generated invoice number is returned by the Desktop Flow; this is also available for the subsequent steps. In the lower part of the preceding figure, you can see how this information is used in a subsequent step to generate the information activity for the sales representative.

Unattended execution (Run mode)

So far, we have started PAD flows manually and were able to observe interactively how the flow ran through the actions. Here, there was a user logged on to the system. This is what is called an **attended execution**. Accordingly, an **unattended execution** runs without the option to look at the execution. This also means that an active user session can't be on the machine that executes the flow. An unattended execution of a flow can only be triggered from a cloud flow. This is possible because we provided all the necessary information for the flow to log in to the machine and start the flow. During execution, a locked Windows user session is created. This session is released when the execution finishes.

To run the previous example in unattended mode, it would only be necessary to make the appropriate setting in the connector for Power Automate and log out of the local Windows session. Of course, this makes sense, especially if no spectators are desired during the execution of the flow – for example, because sensitive data is being processed.

As we have seen in this section, it is possible to seamlessly incorporate Desktop Flows into larger integration scenarios thanks to the connector capabilities in Power Platform, and without having to code a single line. With these capabilities of PAD, local applications are no longer isolated and can provide their capabilities and data to the business, making a valuable contribution.

Another very popular service in the Microsoft Cloud is SharePoint. We'll take a look at its interaction capabilities in the next section.

Managing lists, files, and folders with SharePoint

Microsoft SharePoint continues to be very popular and is the platform for managing team pages, project workspaces, lists, and document libraries within the Microsoft 365 product suite. Often, this platform is also used as a document management system and serves Microsoft Teams as a substructure for managing documents:

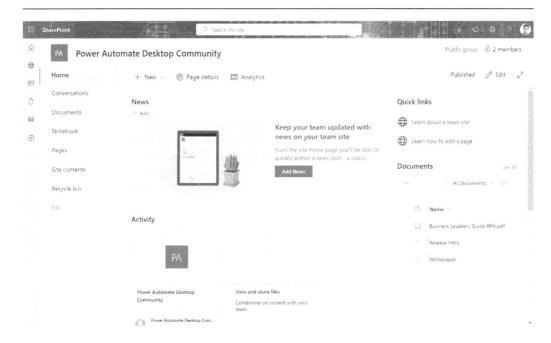

Figure 9.13 – Standard SharePoint example team site

Document libraries contain many useful settings such as versioning, collaborative work on documents, and many more. Therefore, they offer a great advantage if frequently used systems such as SharePoint can also be addressed natively by PAD. Desktop flow creators no longer have to turn to cloud flows to work with SharePoint. All data stored in SharePoint is thus automatically available to a larger team or even an entire company. At the time of writing, the actions for SharePoint are still in the preview stage. Let's take a look at the actions grouped by type:

- **File**:

  - Copy or move a file from one site address to another site address

  - Create or update a (new) file from binary data

  - Check in, check out, or discard a file to prevent someone else from overriding changes

  - Get metadata and delete and get file content by the file identifier

- Folder:

  - Copy or move a folder from one site address to another site address

  - **List folder and list root folder**: Returns files contained in a given folder or the root SharePoint folder

  - **Get folder metadata**: Returns metadata information about the folder

  - **Extract folder**: Extracts the content of a ZIP file to a folder

- **List**:

  - **Get lists, get all lists and libraries**: Returns all lists or both (lists and libraries) from a given SharePoint site

  - **Get list views**: Returns all views from a given SharePoint list

Files and folders should be understood here in the same way as on a local computer. Lists and libraries represent a superordinate container in which files and folders can be contained accordingly. Please note that there are no actions to upload or download a file directly. But this can be accomplished by using additional PAD actions, which is what we will see in this section's example.

SharePoint actions can work with both the online version and a local installation. For the latter, a connection must be established via the Data Gateway.

However, before these actions can be used, PAD still needs a connection reference to SharePoint. We learned about the connector principle in the previous chapter. From this point of view, SharePoint is just another service for which a connector must be set up:

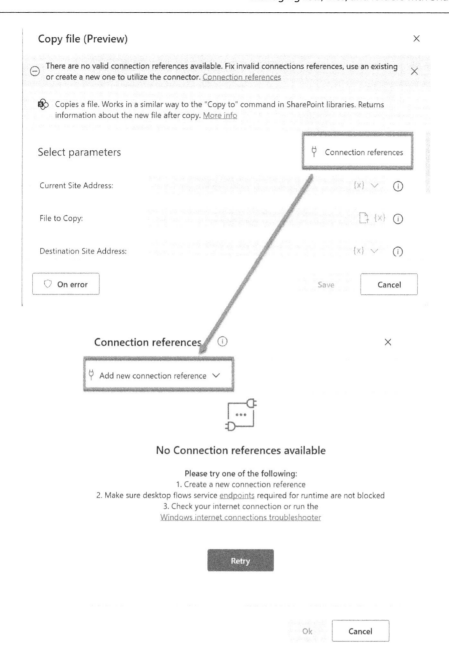

Figure 9.14 – Creating a SharePoint connection and reference

Connections and connection references

To use the SharePoint actions in PAD, we need a connection, as well as a connection reference. Both can be created from within PAD. The concept of connection references was introduced by Microsoft to abstract from a specific connection to an API or service, which can be different in different environments. Additionally, a connection reference can be part of a solution (a transportable unit of customization). All this is just background information and not necessarily important in the context of PAD unless we enter the enterprise area.

As an example, I created a simple page in SharePoint. This already contains lists, as well as a document library. The individual actions should be familiar to SharePoint users and are largely self-explanatory. Only the connection with data from a local computer is not covered by the SharePoint actions alone. Therefore, the following small example shows how a local file can be uploaded to SharePoint via PAD (see the following figure):

Figure 9.15 – Uploading a file to SharePoint via PAD

In this example, an action from the file action group is being used to convert the file in question into binary format and store it in a variable. Then, the content of this variable is used so that I can use the **Create File** action of the SharePoint action group to create the new file in the document library.

In conclusion, the availability of SharePoint actions within PAD will prove to be very handy as it will allow many flows to be created directly in PAD without you having to additionally rely on cloud flows and their connectivity to SharePoint.

In addition to the SaaS services discussed in the previous sections, Microsoft also has IaaS services, just like AWS. What these IaaS offerings are and how these services can be used will be explained in the following section.

Working with IaaS offerings (Azure/AWS)

If we go through the list of PAD actions, we will also come across **Azure** and **AWS**. Therefore, PAD addresses the two largest providers of so-called IaaS providers, namely AWS and Microsoft Azure. Although both providers have other offerings for cloud services (**Platform-as-a-Service** and **Software-as-a-Service**), we can only use a limited area of the cloud offering with the actions available here, namely for controlling virtual machines and storage. The backgrounds for the use of VMs are manifold:

- With PAD, we automate those applications that can only run on a local machine and do not provide any interfaces to the outside. However, this application does not necessarily have to run on the machine where PAD is also executed. As shown in the previous chapters, we can also use Remote Desktop to access other machines and control applications installed there.

- A VM can be set up in such a way that a certain operating system or special components for running the application are also installed, but we do not want to and cannot install them on our PAD machine.

- VMs can be ramped up and down at will. This means that if an application is only needed once a month for a certain workload, for example, a lot of resources can be saved and the machine can only be booted for the time of use.

So, the respective PAD actions are designed to control virtual machines (start, stop, restart). In addition, it is possible to work with virtual disks, which can be created and deleted. Once created, a disk needs to be attached. Please note that we don't have any actions to create or delete a VM. Thus, it is assumed that the applications that are to run on the VM are already installed.

The following table shows the actions that can be used within PAD for managing virtual machines and storage:

	AWS	Azure
Name of the offering	Elastic Compute Cloud (EC2)	Azure Virtual Machines
PAD actions group	AWS and subgroups	Azure and subgroups
Connection via	Create EC2 session and end EC2 session	Create session and End session
VM capabilities		
Start/Stop/Reboot	Start EC2 instance, Stop EC2 instance, Reboot EC2 instance	Start virtual machine, Stop virtual machine, Shutdown virtual machine, Restart virtual machine
Get list of machines	Get available EC2 instances	Get virtual machines
Get description of machines	Describe instances	Describe virtual machine
Snapshot capabilities		
Create/delete	Create Snapshot, Delete Snapshot	Create snapshot, Delete snapshot
Describe	Describe snapshots	Get snapshots
Disk storage capabilities		
Create/delete	Create volume, Delete volume	Create managed disk, Delete disk
Attach/detach	Attach volume, Detach volume	Attach disk, Detach disk
Get information on disks	Describe volumes	Get disks

Table 9.1 – Comparison of actions for VMs

Azure actions are completed with the possibility to work with so-called resource groups (get, create, delete). These provide a logical and thematic container to which Azure artifacts can be assigned to easily identify, for example, all resources that belong to the **order management** topic.

To work with these actions for each of the providers, a connection must be established via the corresponding **session** actions. The connection setup is handled differently for the two providers.

AWS actions in use

The AWS session action is depicted in the following screenshot. There are two ways to create a session: with a profile name and region or with an access key, password, and region. The latter method, **Access keys**, is shown here:

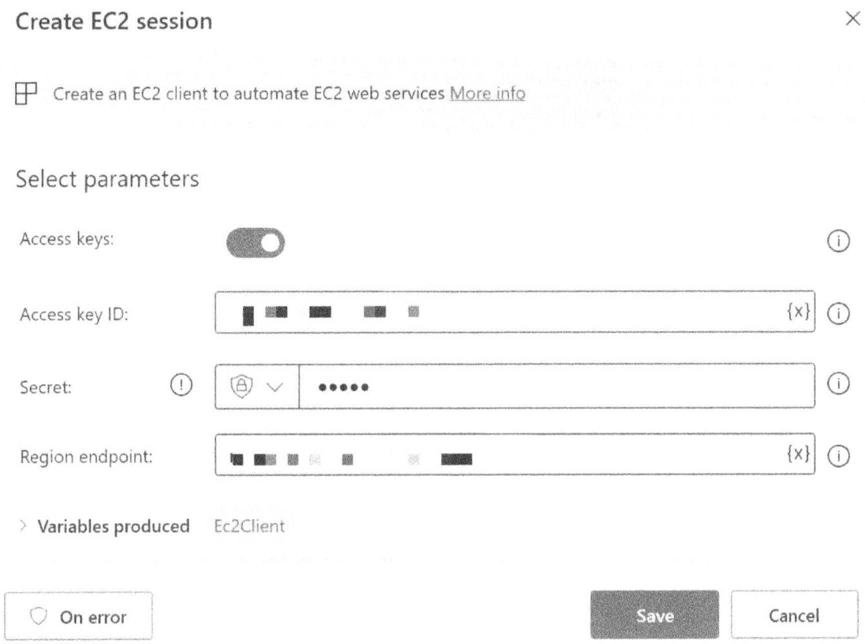

Create EC2 session   ×

⊞ Create an EC2 client to automate EC2 web services More info

Select parameters

Access keys:	⬤		ⓘ
Access key ID:	▮ ▬ ▬ ▬ ▯	{x}	ⓘ
Secret:	ⓘ	🛡 ∨	•••••	ⓘ
Region endpoint:	▬ ▬ ▬ ▮ ▮ ▬	{x}	ⓘ

> **Variables produced**   Ec2Client

♡ On error   **Save**   Cancel

Figure 9.16 – Creating an AWS session action

If a trial version of AWS already exists, the first step is to create an access key for the service. After logging into the main portal at https://aws.amazon.com, it is recommended to select the correct region in which to work and create the virtual machines. As shown in the following figure, Ireland has been selected here:

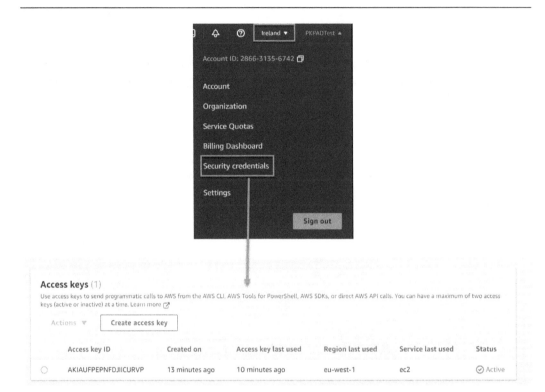

Figure 9.17 – AWS access key and secret

Clicking on the **Security credentials** link opens the **Identity and Access Management** page. A little further down this page is a section for access keys. Here, you can easily generate a new access key. The password (or secret) is only displayed once. It is advisable to copy this password to another location. Depending on the selected region, the entry for this parameter must still be made. A list of the available service endpoints and their entries can be found in the *Further reading* section. For **Ireland**, which we have selected here, the string is `eu-west-1`.

With these three entries, an EC2 session can now be successfully created and thus all further actions can be used.

For the following demo UI flow to work, we also need a machine provisioned in EC2. There are numerous guides on how to create VMs in EC2. The test account can be used to create both Linux and Windows VMs, which are then managed via the console. For this flow, a simple Windows machine was created.

The UI flow shown in the following figure logs into AWS with the credentials (**1**), then calls the available EC2 instances and stores them in a variable (**2**).

The subsequent loop goes through all instances (**3**) and checks each one to see whether it is stopped or not (**4**). If stopped, a message box asks whether the machine should be started (**5**). This is what the flow looks like:

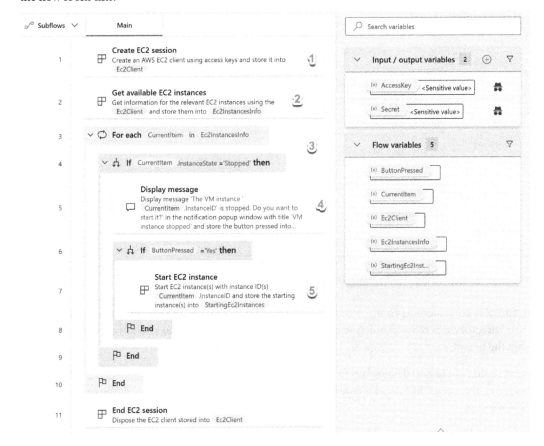

Figure 9.18 – Working with AWS EC2 instances

Working with Azure actions

In the first section of this chapter, we saw that Azure is the foundation for all Microsoft Cloud services. To explain the concept of Azure without going too much into the details, take a look at the following diagram:

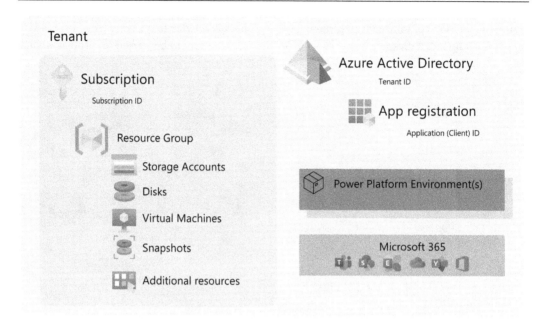

Figure 9.19 – Azure relationship between different components

The tenant is equivalent to an organization or a company. Typically, the tenant contains an Azure Active Directory where we can define and register new applications that need to access resources within the tenant.

A subscription is a financial boundary for any kind of resource. Because resources consume CPU and/or memory, there must be financial coverage of the consumption. There can be multiple subscriptions within a tenant.

The next level is a resource group, which usually relates to a specific purpose or topic. For example, it could be possible to create a resource group that contains all resources for automating a specific workload or application. These resources can be storage (disk or volume to persist data) or virtual machines and many more types of resources.

For PAD to be able to access resources in Azure, we need to register a new app within Azure Active Directory. This would have a new client ID as a result, which we can use for the corresponding field in the **Actions Parameter** dialog. Within this app registration, it is also possible to create a client secret, which is a fixed password to use for logging in with this client. In the Azure portal, go to **Azure Active Directory | App registrations** and create a new registration:

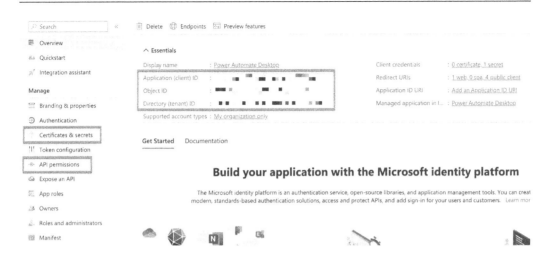

Figure 9.20 – Azure Active Directory – App registrations

Together with the **Tenant ID** (or **Directory (tenant) ID** in the preceding screenshot) and **Subscription ID** properties (needs to be looked up in the subscription), we have all the information for the Azure session to log on:

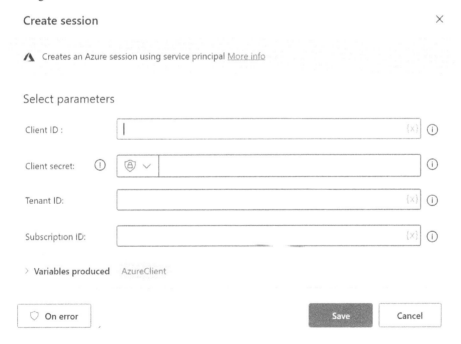

Figure 9.21 – Creating a connection to Azure

> **Configuring additional security settings**
>
> For the Azure session to work, additional security-related steps must be taken. First, the app registration must be granted API permissions for Azure resources to access. These include permission for **Azure Service Management** and **Azure Storage**. Second, the app registration must be given additional role assignments at the subscription level. For this, go to the subscription management in the Azure portal, switch to the **Access Control (IAM)** area, select the **Role Assignment** tab, and use the **+Add** button to start the wizard. Select the **Virtual Machine Contributor** role and add the app registration that was previously created as a member of the role assignment.

The following example shows how to work with actions for Azure virtual machines and disks:

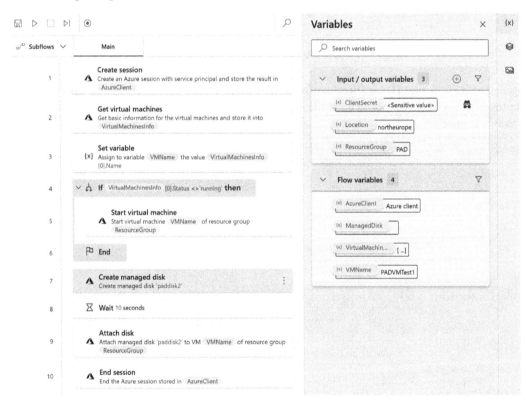

Figure 9.22 – Working with Azure actions

As mentioned earlier, we don't have any actions to create VMs in Azure. This example assumes that a VM has been created in a specific resource group. In *step 1*, the Azure session is created. *Step 2* queries the subscription for any virtual machines. It is also possible to filter this query down to machines in a specific resource group. In this case, only one machine is registered, whose name is assigned to a variable in *step 3*. *Steps 4* to *6* check whether the VM has been started, and if not, the machine is started. Note how the individual properties for a VM can be used to determine the name or status. Next, a virtual hard disk (disk) is created. Various parameters are also required for this object, such as the region, the resource group, and, of course, the name. Since the process of creating the disk is asynchronous in Azure, it cannot be assumed that the disk will be available immediately when this action is completed in *step 7*. Therefore, the flow is paused for 10 seconds in *step 8* to complete the disk creation in the background. Finally, the new disk is attached to the VM and is available directly in **Disk Management** as an unformatted drive.

As a consequence, other scenarios can be feasible. For example, it would be possible to fill a new disk with data from a snapshot or another storage account.

Furthermore, the VM could also be equipped with an agent for virtual desktops so that UI elements can be identified and used in a flow.

In this section, we learned that we can use IaaS providers such as Microsoft or AWS via the native PAD actions to move workloads from on-premises machines to VMs and then control them through a locally executed UI flow.

This can save resources, especially if an application is not used for automation daily or uses a lot of compute capacity (CPU/memory).

Summary

Through integration with Power Platform, PAD can close the gap between workflows in the cloud and the local network. We have seen that this can create a seamless overall concept for an automation solution, as the various components are very well integrated.

Additionally, we learned about native actions for using IaaS offerings from Microsoft and AWS, which allow us to outsource the execution of desktop programs or add more resources to our flow as needed.

In the next chapter, we'll look at how we can incorporate artificial intelligence into a UI flow while learning about offerings from various vendors.

Further reading

Further information is listed here:

- Sign up for a free Dynamics 365 Sales trial: `https://learn.microsoft.com/en-us/dynamics365/sales/sign-up-for-sales-trial`

- Power Platform environments overview: `https://learn.microsoft.com/en-us/power-platform/admin/environments-overview`

- Connection versus Connection references: `https://learn.microsoft.com/en-us/power-apps/maker/data-platform/create-connection-reference`

- Introduction to Amazon EC2: `https://docs.aws.amazon.com/AWSEC2/latest/UserGuide/concepts.html`

- Amazon Service endpoints: `https://docs.aws.amazon.com/general/latest/gr/rande.html`

- Get started with Amazon EC2 Windows instances: `https://docs.aws.amazon.com/AWSEC2/latest/WindowsGuide/EC2_GetStarted.html`

- Register an application with the Microsoft identity platform: `https://learn.microsoft.com/en-us/azure/active-directory/develop/quickstart-register-app`

10

Leveraging Artificial Intelligence

The topic of **artificial intelligence** (**AI**) has been hotly debated for some time and has gained further popularity with the advent of OpenAI and ChatGPT. In this chapter, we want to shed light on what possibilities we have within PAD in this regard and look at the following areas:

- An introduction to cognitive services
- Diving into the offerings of Microsoft, Google, and IBM
- Classification in the practical work with PADT

After reading this chapter, we will have a solid understanding of what AI means, in which areas this technology helps optimize business processes, and how these functionalities can be deployed and used with PAD.

Technical requirements

The services addressed in this chapter are accessed online by the respective providers. This requires a corresponding subscription with the vendors, who all provide a test period with certain credits free of charge. The process for starting a trial and subsequent steps are different in each case and are highlighted in the corresponding sections.

Introduction to cognitive services and artificial intelligence

The functionalities discussed in this chapter all deal with AI and **machine learning** (**ML**). Before we turn to the individual providers and their functionality, we will first classify the topic in general and consider what benefits AI can bring to everyday life and in the business context.

In general, it can be said that AI can help to recognize structures in text, image, or sound information so that they become machine-readable. Cognitive services hereby refer to a set of cloud-based **application programming interfaces** (**APIs**) and tools provided by Microsoft, IBM, Google, and other vendors that enable developers and creators to add advanced AI capabilities to their applications without needing to have deep expertise in ML or data science. The actions offered in PAD by the different vendors provide more or less the same functionalities, shown in the following diagram. Only the naming of the services is slightly different in some cases.

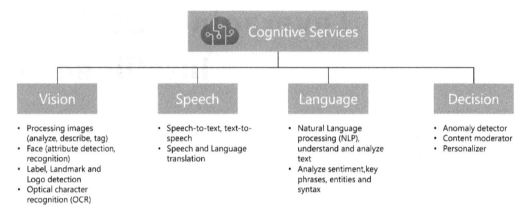

Figure 10.1 – An overview of cognitive services

Let's take a slightly closer look at these capabilities and how they can be used in a business context.

An overview of cognitive services

Cognitive services can also be described as a set of APIs and tools that use ML algorithms to enable computers to perform tasks that normally require human-level intelligence. These services are typically hosted in the cloud and can be accessed through a REST API or **software development kit** (**SDK**). The technology behind cognitive services can be broken down into a few key components:

- **ML**: At the heart of cognitive services is ML, a subfield of AI that involves training computers to recognize patterns in data. The ML models used by cognitive services are typically trained on large amounts of labeled data and are designed to improve their accuracy over time as they are exposed to new data. The latest ML models start already by 100 datasets and are still improving.

- **Deep learning**: Many cognitive services also use deep learning, a type of ML that involves training artificial neural networks with multiple layers. Deep learning models are particularly effective for image and speech recognition, as well as **natural language processing** (**NLP**).

- **NLP**: Some cognitive services, such as language translation and text-to-speech, use NLP algorithms to analyze and understand human language. NLP involves parsing text into its constituent parts, such as words and sentences, and using statistical models to identify patterns and relationships between these parts. **Language Understanding (LUIS)**, for example, is an NLP service that enables developers to build chatbots and other conversational interfaces.

- **Computer vision**: Computer vision is a branch of AI that involves training computers to recognize and interpret visual information from images and videos. Cognitive services that use computer vision can perform object detection, facial recognition, and image classification tasks.

- **Cloud computing**: Finally, cognitive services are typically hosted in the cloud, which enables them to be accessed by developers from anywhere in the world. Cloud computing also provides scalable, on-demand computing resources, which can help to reduce costs and improve performance.

Overall, cognitive services are powered by a combination of advanced ML, deep learning, NLP, and computer vision algorithms, made accessible through cloud computing technologies. These services can help to automate many tasks that would otherwise require human-level intelligence, such as speech recognition, language translation, and image analysis. Cognitive services make it easier for people to add powerful AI capabilities to their applications without needing to have extensive expertise in ML or data science.

AI in the context of business

But how can these skills be applied meaningfully in a business context? Here are some examples of how cognitive services and image analysis can be used in different business scenarios:

- **Customer service**: Chatbots powered by cognitive services can be used to provide customer service 24/7, answering frequently asked questions, helping customers find products and services, and resolving simple issues. These chatbots can be integrated into websites, messaging apps, and social media platforms. By including sentiment analysis, such chatbots can be used to analyze customer feedback and sentiment on social media platforms and review sites. This can help businesses to identify issues and address them quickly, as well as identify areas where they are excelling. Cognitive services can be used to power voice assistants that can be used for customer service. For example, a customer might use a voice assistant to track an order, update their account information, or ask for product recommendations. Cognitive services can be used to automatically create support tickets based on customer inquiries received via email, chat, or social media. This can help customer service teams to more efficiently manage their workload and respond to customers in a timely manner.

- **E-commerce**: Cognitive services can be used to analyze images of products and recommend similar products to customers based on their preferences (upselling and cross-selling). For example, a customer might take a picture of a dress they like, and an e-commerce site could use image recognition to suggest similar dresses in different styles or colors. Cognitive services can be used to analyze images of products that customers submit as part of their support requests. This can help customer service representatives quickly understand the issue and provide more accurate and efficient support. Cognitive services can be used to detect fraudulent activities in e-commerce, such as the use of fake or stolen product images. This can help to protect the business and ensure that customers are getting what they paid for.

Basically, computers can only process machine-readable information. Optical character recognition is a technique that has been around for a very long time and can help with this very issue. Let's take a look at what it's all about.

Capabilities with optical character recognition

It was recognized very early on that although computers can also handle images very well, they lack the ability to recognize text in images and thus to further process the information in them. This problem is solved by **optical character recognition** (**OCR**). Such OCR technology has been implemented programmatically without the use of AI.

> **Legacy OCR actions**
>
> There is also a separate action group, OCR, in the list of action groups in addition to the actions within the context of cognitive services. These actions in OCR rely on locally installed engines that come with PAD and are not based on AI, and will soon be discontinued and no longer supported.

With the availability of higher computing capacity, OCR can now be performed with AI's support. This technology can recognize and further process printed or handwritten text that may be located on various objects in reality. With this, it is possible to turn text in an image into machine-readable text. Moreover, with additional functions such as text analytics, it is possible to grasp the meaning of a text, create summaries, and make connections.

All cognitive services providers have the following in common:

- A corresponding account at the provider is required to use the services

- All requests need an API key to authorize the request

- All results come with a certain level of confidence and a status code as a response

Let's take a look at what each vendor has to offer for AI within PAD.

Exploring Microsoft Cognitive Services

To make use of the actions for Microsoft Cognitive Services, we need to have an Azure subscription and provision a resource called **Cognitive services multi-service account**. In the following screenshot, you can see a ready-configured service:

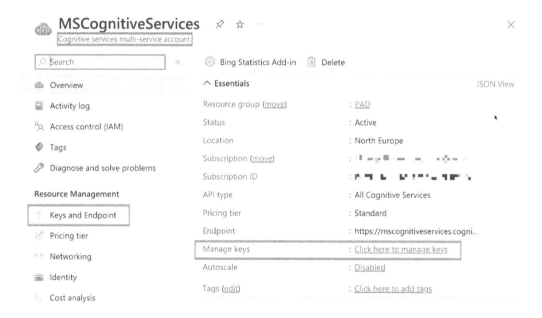

Figure 10.2 – Microsoft Cognitive Services in Azure

Setting up this resource does not require complex settings. After this, the actions for Microsoft Cognitive can be used. Each of these actions expects a **Subscription** key. However, this is not the ID of the subscription itself, but the key that can be retrieved in the previous screenshot under **Keys and Endpoint**. Simply call this page and copy one of the pre-generated keys, and set it as the value for the Subscription Key parameter (or save the value as a variable and set it). Now let's take a look at what intelligent services are available from Microsoft within PAD:

- **Bing spell check:** The Spell check checks a search term for syntactical and colloquial errors and returns a corrected version. This is used to submit better and more accurate search queries and thus get more and better search results.

> **Hint**
>
> The Bing News API has been removed from Cognitive Services.

- **Text Analytics**: This allows us to provide some text and then detect the language, extract key phrases, and determine the sentiment of the text. This allows conclusions to be drawn about the text and whether the author was in a good mood in relation to the content.

- **Computer Vision**: This contains the actions for image analysis and description as well as the AI-based version of OCR. The image-related actions work with an image that can be provided locally or via a URL. The actions are as follows:

  - **Analyze image**: This can extract a variety of visual features from the image, for example, specific brands, objects, or human faces.

  - **Describe image**: This returns a list of attributes that describe the image.

  - **Tag image**: This generates content tags for an image, each with a separate confidence level.

  - The **OCR** action can extract text from an image. This includes posters, street signs, and product labels and also documents such as articles or invoices. The result of this action is a complex object with a list of paragraphs containing a list of text lines containing a list of words.

Let's take a look at an example of how to use the actions in Computer Vision within a **user interface** (**UI**) flow.

AI-based image sorting

Imagine you have called a contest to choose the most beautiful pet by species from all senders of a certain group. Now, you receive countless emails with lots of pictures and put them in a folder on your desktop. We can now use AI to sort the images by pet type and create individual folders for each. The corresponding UI flow for this looks like this:

Figure 10.3 – Using Computer Vision in PAD

As a prerequisite, I created a base folder on my desktop containing some random pictures of dogs and cats. I created an `BaseFolder` input variable that contains the location of that folder. Here is what this flow does:

- In step 1, a collection of the pictures is created and stored in the `Files` variable.

- Step 2 starts the loop to iterate through the list of images.

- In steps 3 and 4, I inserted the **Describe image** and **Tag image** actions for illustrative reasons.

- The **Analyze image** action is called in step 5. This will store the identified content in the corresponding `JSONResponse` variable evaluated in step 7.

- Step 6 is the beginning of a block for error handling. If the analysis of the image does not return an identified object, the `JSONResponse` variable will be empty. Because we don't want the flow to stop processing, we need to encapsulate the following actions in that error block.

- Step 7 evaluates the response and stores the category in a separate variable. The category essentially contains the animal species.

- In step 8, we are checking whether a folder for that species exists and creating one if necessary (step 9).

- Step 11 moves the current image into that folder so that the base folder gets cleaned up.

After these flows have run, all images that have been analyzed are located in dedicated subfolders. To also demonstrate the difference between the Computer Vision actions, here is a screenshot of the results for `ImageTags`, `ImageDescription`, and image analysis:

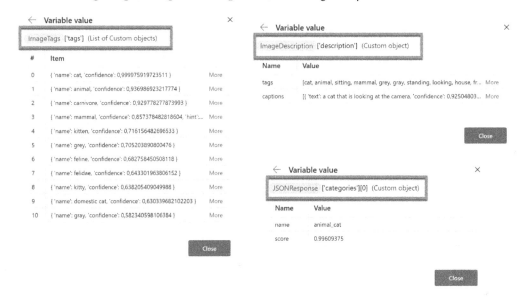

Figure 10.4 – Image tags, description, and analysis

Please note the **captions** entry in `ImageDescription`, **a cat that is looking at the camera**, which shows us how powerful AI can be. Let's look at the next vendor and the actions that it provides.

Google cognitive

Google's AI functionalities are accessible via the **Google Cloud Platform** (**GCP**). First, you need to create a corresponding account. The provider provides a certain allowance that can be used to try out the services. The entry is carried out via `https://cloud.google.com`, where a new account can also be created. You will then be in the Cloud Console, where you can search for resources and also create a new project to work with those APIs. For example, you could use the **Search** box and look for the term `sentiment`, which is part of the Cloud Natural Language API. Each of the APIs needs to be enabled separately. You can take a look at all enabled APIs from the burger menu at the top left of the screen, as can be seen in the following screenshot:

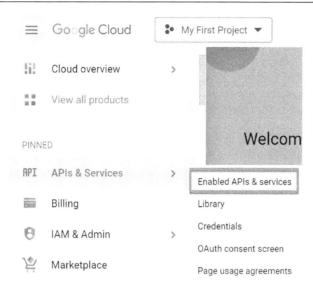

Figure 10.5 – The Google Cloud Console

After an API has been activated, credentials must still be created to be able to call this interface from outside. This also applies to the actions in PAD, as these require an API key for execution. For each service, there is a **Credentials** section, which can be used to create an API key (see the following screenshot).

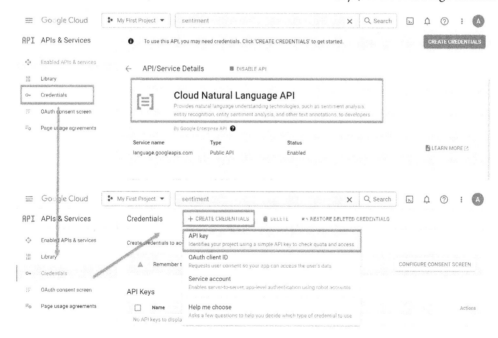

Figure 10.6 – Create an API key in Google Cloud

After this, **API key** is displayed in a window so that it can be copied and pasted into one of the actions for PAD. This process must be repeated for each API that is to be used. Two APIs can be used within PAD, Natural Language and Vision:

- **Natural Language**: This allows us to analyze the sentiment (the mood in this text), entities (what the text is about), or syntax (how the text is structured) of a given text.

- **Vision**: This takes an image and can detect different objects such as labels (or objects), landmarks, text, logos, and image properties. It also allows to detect saved search.

Let's take a look at a very simple UI flow to demonstrate these results.

Sentiment analysis in PAD

In this example, we take a text document with some content:

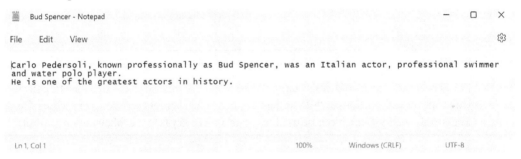

Figure 10.7 – A sample text file for AI analysis

We can now create a flow and drag the **Analyze sentiment** and **Analyze entities** actions into the designer, equipped with an API key and the location of this file.

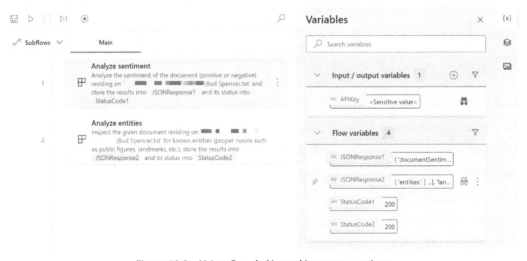

Figure 10.8 – Using Google Natural Language actions

The result of the sentiment analysis looks like this (double-click on the `JSONResponse1` variable):

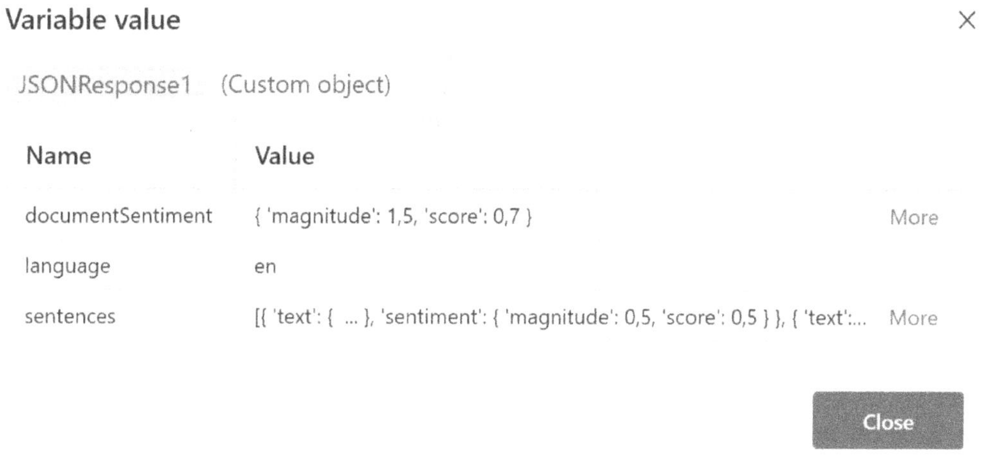

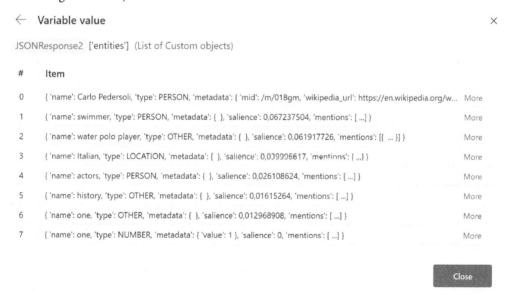

Figure 10.9 - The result of the sentiment analysis

We can see that we receive a scoring and a magnitude for the overall sentiment of the document. Additionally, each sentence is also evaluated separately for the sentiment. According to the documentation, a score of about 0.0 would be neutral, as well as a magnitude with a low value. This would allow a slightly positive sentiment as an interpretation.

The result of the second **Analyze entities** action gives us some insight into what this text is about (see the following screenshot):

Figure 10.10 – The result of entity analysis

Here, I expanded the `entities` branch by clicking on the **More** link in the dialog. We can see that the person's name was recognized, even with a link to Wikipedia, and also other entities in the text.

All these functions may not be particularly helpful on their own at first. However, when viewed in a larger context, important and helpful insights and automatisms can be gained through these functionalities. For example, by extracting this information and intelligently connecting it afterward, summaries of longer texts can also be created. This inevitably also leads companies to create very large language models that can analyze and understand texts and also provide intelligent answers. This can be seen, among other things, in the increasing popularity of models such as ChatGPT.

Working with IBM cognitive

Maybe some of you remember the headlines from 1997, which talked about a computer program called Deep Blue, which won the world championship against Garry Kasparov, the reigning world chess champion, in only 19 moves. The program ran on a supercomputer built by IBM. The victory against the world chess champion was a milestone in the history of AI. Today, IBM offers its AI-based solutions under the name Watson. As mentioned earlier in the chapter, the functionalities of the individual providers overlap. IBM provides the following functionalities:

- **Visual recognition**: This action group contains a **Classify image** action, which allows us to provide an image
- **Language translator**: This allows us to identify the language and translate a given text
- **Document conversion**: This converts a given document into another format

The **Tone analyzer** action group contains an **Analyze tone** action, which is deprecated by now and will not be considered further here. If there are already investments in IBM Cloud with AI, these offers can also be used in the actions available in PAD.

As with the other providers, an account for IBM Cloud is also required here, via which a corresponding API key can then be generated. This information must also be specified for the PAD actions to execute the request. Otherwise, the basic functionality is exactly the same as with the providers shown so far.

Now that we have learned about the functionalities of the different providers, let's take a look at the context in which the use of these capabilities makes sense.

Practical relevance to PAD

The actions provided under PAD are not unique to PAD, other than launching a local application and controlling it. All APIs called by these actions are available in the cloud and, therefore, available to any application.

The reason we see these actions in PAD as well is that it gives us direct access within the context of UI flows without the need to add more cloud flows or other technologies. This allows us to build some intelligence into our UI flows. In selected scenarios, these actions can certainly make sense, but typically, this functionality will be used primarily in cloud scenarios. However, in cases where the content of an image is to be determined within PAD or the mood in an email is to be determined, the availability of these actions is, of course, very helpful. The availability of these actions can be helpful, all of which rely on pre-built and trained AI models. Another way of using AI that is more adapted to a business is described in the following section.

More ways to use AI

AI and its capabilities have gained popularity in recent years, and we can see an increased use of this technology. In Power Platform and Dynamics 365 alone, AI is used in many places, such as the following:

- **Conversational intelligence**: Information and insights are extracted from a seller's call recordings by using analytics and data science to proactively coach sellers.

- **Intelligent content suggestion**: Email responses can be suggested based on the previous conversation. Questions can be detected, and possible answers suggested, which helps a seller to be very quick and productive.

- **Power Virtual Agent**: The intelligent chatbot within the Power Platform can leverage LUIS models to digest the meaning of conversations in chats and provide corresponding responses.

- **Copilot for Power Automate**: This allows the creation of a Power Automate by verbal and natural language description (see `https://powerautomate.microsoft.com/en-us/blog/new-ways-to-innovate-with-ai-and-microsoft-power-automate/`).

However, there is one scenario that I would like to discuss briefly here, which could certainly have relevance for practical use. There are always situations where certain paper-based forms need to be processed in day-to-day business:

- The school form where parents are supposed to tick certain things to give their consent or fill in the information

- The order confirmation or delivery bill from a supplier with transaction data

- The form at the hospital or doctor's office for patient registration

The basic mechanism is always the same, as there is always a specific form on which data is always entered in the same place. Microsoft and Google, for example, offer prefabricated but not yet trained models for digitizing and reading out these forms. The Microsoft Power Platform contains such a service, which is called AI Builder. Here, we find predefined models for the following purposes:

- Extracting data from PDF documents, identity documents, business cards, and so on

- Classifying customer feedback or any other text into predefined categories

- Extracting key elements from text or the most relevant words and phrases from text

If these models are not sufficient, it is also possible to create your own AI models. After the models have been equipped with training and test data, they can be deployed directly and used within a cloud flow. Accordingly, this technology allows you to use your own forms with AI and incorporate them into a process.

In principle, such a model would also be retrievable via a standard HTTP request. In the following chapter, we will learn how this and other web protocols can be used with PAD.

Summary

In this chapter, we looked at the topic of AI and the options for adding intelligence to UI flows. We learned that these options usually play a role in a larger context and that, within PAD, we also have access to the common models today. However, if we find the need to access such functionality as part of a UI flow, we can do so within PAD using the actions presented here, without having to consider other solutions and technologies.

Further reading

- *Microsoft Computer Vision documentation*: https://learn.microsoft.com/en-gb/azure/cognitive-services/computer-vision/

- *Deep Blue versus Kasparov*: https://en.wikipedia.org/wiki/Deep_Blue_versus_Kasparov,_1997,_Game_6

- *Google Natural Language AI documentation*: https://cloud.google.com/natural-language

- *Google Vision AI documentation*: https://cloud.google.com/vision

11

Working with APIs and Services

So far, we have used PAD to enable applications without an API to be used with automated processes via their **user interface** (**UI**). Even though this is the main purpose of PAD, we can also communicate with APIs in the context of UI flows to wherever is the best integration option. These possibilities close the loop and allow us to implement a complete business process exclusively in PAD without adding other mechanisms. Accordingly, in this chapter, we will look at which APIs and protocols PAD can communicate with directly and which web-based data formats PAD can handle. We will highlight the following points:

- Managing local **Active Directory** (**AD**) services and accessing local database and email services
- Leveraging web services with **Hypertext Transfer Protocol** (**HTTP**) and **File Transfer Protocol** (**FTP**) and working with web formats (**Javascript object notation** (**JSON**), **Extensible Markup Language** (**XML**), and **portable document format** (**PDF**))
- Cryptography and CyberArk
- Built-in data functionality – date-time, text, compression, and Messageboxes

After studying this chapter, we will be able to use internal services such as databases and AD, as well as any web services within a UI flow, working with the various data formats. An example flow for the actions presented here will be shown in *Chapter 12*.

Technical requirements

In the first example, I will use a locally installed SQL Server with the Microsoft sample database **AdventureWorks2019** installed. In addition, I have set up a database user on this server with access permission to this database.

The use and automation of local network services

In many large and small companies, Windows-based local networks are used, providing additional Microsoft services. First and foremost, of course, this includes AD, but also Microsoft Exchange, SQL Server, and so on. These and other services are assigned to the Office Server product line.

Since PAD is perfectly suited to automate administrative tasks, it is obvious that corresponding actions are also provided for these services. Unlike the local Windows services we learned about in *Chapter 7*, these are network services on the local network and, therefore, not in the cloud. Accordingly, the actions presented here are also to be used only for the on-premises installed server products. To automate the corresponding cloud services, Microsoft refers to the use of the respective connectors for the online products (Azure AD, Exchange Online, and so on). The fact that more and more companies are opting for cloud technologies is probably also due to the fact that the supported server version for Microsoft Exchange is somewhat older (Microsoft Exchange Server 2010-2013 Service Pack 1). While the actions for AD allow a limited amount of administration, the actions for Exchange and Database can only be used to use these services, for instance, retrieve messages or execute SQL commands. Let's look at the details of the actions.

Accessing and managing local AD

AD is based on the **lightweight directory access protocol** (**LDAP**), which is based on a simpler subset of standards contained in the X.500 standard. Microsoft AD is the central system in local Windows networks to manage network objects such as users and groups, computers, printers, and more.

Background information on AD services

Microsoft AD consists of a set of services such as **Federation Services** (**FS**), **Certificate Services** (**CS**), and also **Domain Services** (**DS**). Among the various AD services, the so-called DS are meant here, which can be interacted with via the PAD actions. These are the foundation of a Windows domain network.

When working with AD and the corresponding PAD actions, you will typically come across these terms:

- **Object**: This is the general term for every item that is managed within AD. Each different object type has specific attributes.

- **User**: This object type consists of attributes such as the user's logon name, first name, and last name, and represents someone logging on to a Windows domain with this user account.

- **Group**: This consists of user accounts, other groups, or computer accounts.

- **Computer**: This is information that relates to a computer within a domain.

- **Organizational unit** (**OU**): These are logical containers that can contain objects and also other OUs. Often, OUs are used to mirror the organization's structure. They also allow the delegation of administrative control.

AD is managed through management tools that come with Windows Server. The following screenshot shows the **Active Directory Users and Computers** console used to manage the objects described in the previous list.

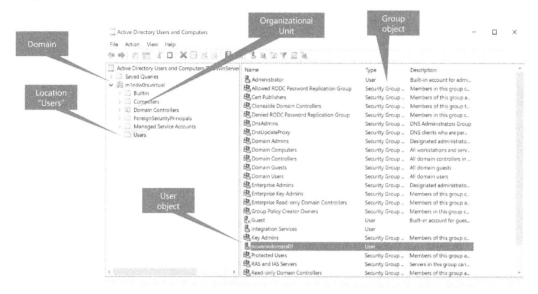

Figure 11.1 – AD management with Server tools

Another important concept is the distinguished name. This name consists of the domain name, the location, and the object's name. The combination of those entries must be unique, which is also the reason why there cannot be two user accounts with the exact same name. The distinguished name is a special syntax to describe objects in AD. In the preceding screenshot, we see a user account called **powerautomate01** in the Users file location. The distinguished name for this entry would be CN=powerautomate01,CN=Users,DC=m1ndw0rx,DC=virtual.

> **Pro tip – viewing the distinguished name in the management console**
>
> The management console shown previously is also able to expose the distinguished name for any object. To see this, you need to turn on the **Advanced Features** option in the console by clicking on the **View** menu and **Advanced Features**. This results in some more entries in the list of objects but also in additional tabs in the **Properties** dialog of an object. Right-click on a user account now and select **Properties** from the context menu. You will see an additional tab called **Attribute Editor**. Scroll down the list until you find the distinguishedName attribute and double-click it to see the entry.

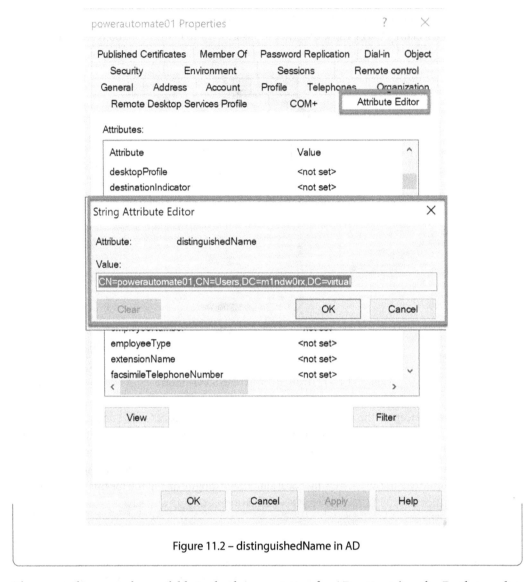

Figure 11.2 – distinguishedName in AD

There are other examples available in the documentation for AD actions (see the *Further reading* section). Now, let's use this knowledge to see what PAD offers in terms of actions to manage AD:

- **Connect to server / Close connection:** This establishes a connection to an AD server by providing an LDAP path to the directory, the fully qualified server name (including domain information), or even the IP address. There is also an option to provide a username and password if the current user does not have the privilege to access the service.

- **Create / Modify group**: The **Create** action creates a group by providing the name, a location, the group scope (global, local, or universal), and the group type (security or distribution). The `location` attribute specifies the container/folder where the group should be created as a distinguished name. The **Modify** action allows you to rename or delete a group and also to add or remove users from it.

- **Get group info / Get group members**: This provides a distinguished name for the group you wish to get information about. The first action returns a list of attributes for that group, and the second a list of group members.

- **Create / Delete / Move / Rename object**: An AD object in this case is a *computer* or an *OU*. Again, we need to provide a location (distinguished name) to execute these actions.

- **Create / Modify / Unlock user**: The **Create** action also wants a location where the user should be created, but also common attributes such as their first name, initials, last name, username, and password. It also allows you to specify whether the password should never expire or whether the user account will be disabled (the user cannot log on).

- **Get user info / Update user info**: This provides a distinguished name for the user to retrieve or update user object attributes.

Tip – managing AD without membership

Typically, workstations are part of AD, meaning that they are joined to that domain, and users from that domain can log on. The interactive user who also runs PAD could be such a user. Creating a connection to the AD in this scenario shouldn't be a problem. As mentioned previously, it is possible to create a connection to an AD via the IP address of the domain controller and appropriate user credentials. In this case, the computer running that UI flow does not need to be a member of that domain, and a connection could be established via the IP address. However, to use the actions and, for example, create a user, some knowledge of the domain is required to provide the information for the distinguished name.

In *Chapter 8*, we discussed the scenario of new hires that come into the company and therefore need a SAP logon. We created a UI flow that reads through an Excel list and created a SAP user for each entry. As an additional task, we could also create a user in AD in the same way. I want to encourage you to take this example and add this additional task as a sub-flow. If there is no SAP system at hand, it is also possible to take the UI flow from *Chapter 8* and replace the SAP part with a part for AD.

Although there are certainly still a lot of local Windows networks, the fact that PAD includes management actions for a local AD shows where the software comes from and how long it has been on the market. Apart from AD, no other actions are offered that can manage local Windows network services. Nevertheless, the use of the following services also represents great value for PAD as a whole.

Working with databases

The **Database** actions group only contains three actions (**Open SQL connection, Execute SQL statement**, and **Close SQL connection**), but these open up a whole range of possibilities since all databases can be accessed to issue any kind of SQL commands. Until now, the focus has been on systems that do not have any API. However, PAD is also able to integrate systems with an API, so that within a UI flow tasks can also be done, which require the addressing of an API. The key to this universe of data is revealed when dragging the **Open SQL connection** action onto the designer on the right side of the connection string parameter (see the following screenshot).

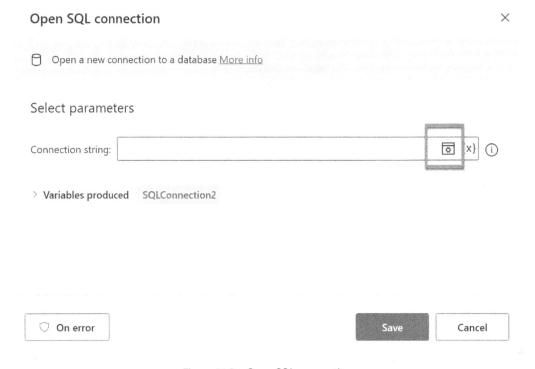

Figure 11.3 – Open SQL connection

To use a database, we need to provide a connection string, and clicking on the red framed icon shown in the previous screenshot opens a wizard to do exactly this.

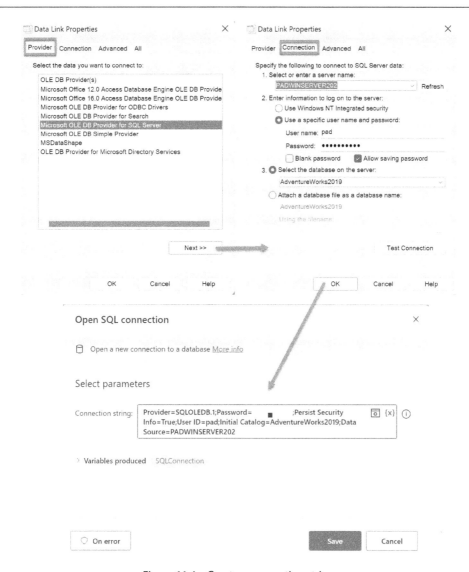

Figure 11.4 – Create a connection string

In my example, I am using a SQL Server database on a remote server, but as you can see in the preceding screenshot, the **Provider** tab contains a lot of entries that can be used to connect other systems. Pressing the **Next** button here takes us to the **Connection** tab, where additional entries have to be made depending on the provider that was selected on the previous screen. I have entered the remote server name and credentials that have been provided, which enables me to select a database on that server. If I now press the **OK** button, the connection string is generated for me and entered into the corresponding text field, where I could also copy the entry and enter it into a sensitive variable.

With this short preparation, I can now execute any SQL statement against the database. Note, however, that statements may not be executed successfully if the corresponding user does not have the necessary permissions. The following screenshot shows the **Execute SQL statement** action. The query result will be stored in a variable for further processing.

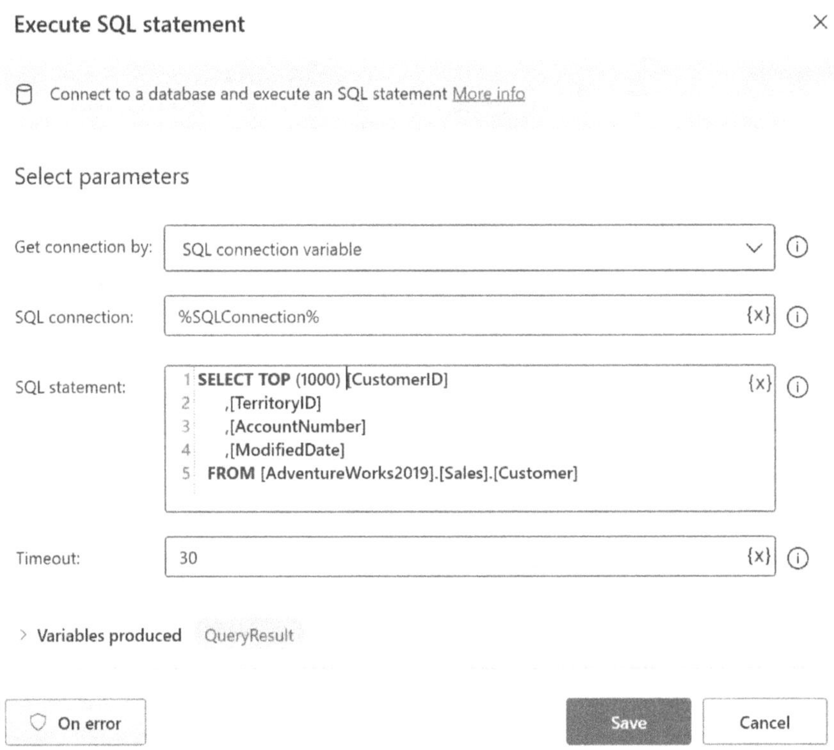

Figure 11.5 – Execute SQL statement

The following comments can be made about the SQL statement:

- Very often, the quality of the statement is crucial for the performance of the database and, thus, the entire system. Therefore, queries such as SELECT * FROM .... should be avoided if possible.

- Of course, other content from previous flow steps that are stored in variables can now be inserted into the statement.

In the next main section, we will look at an example of how a database can be integrated into a larger context. Before that, let's take a look at the other services in this category.

Using local exchange and other email services

We have already learned in *Chapter 2* and *Chapter 8* how to work with a locally installed Outlook application and the corresponding PAD actions to receive, process, and send emails. In addition to this possibility, PAD provides other action groups to work with emails. These behave very similarly to what we have already learned about before, except for the connection setup. These action groups are **Exchange Server** and **Email**. To provide an overview and to see the difference between those action groups, including the one for **Outlook**, let's look at the following table.

Feature/Action group	Outlook	Exchange Server	Email
Connection through	Local Outlook client	Local/on-premises Exchange Server deployment	Any Internet Message Access Protocol (IMAP) mailbox
Connection capability	Any mailbox that is connected through Outlook, including Outlook 365	One personal or shared mailbox	Any IMAP server
Retrieve email messages	Yes	Yes	Yes
Send email messages	Yes	Yes	Yes
Process email messages	Yes	Yes	Yes
Save messages	Yes	No	No
Respond to messages	Yes	No	No

Table 11.1 – A comparison between different email options in PAD

As we can see here, there are many similarities between the different action groups and one major difference. While the **Exchange Server** and **Email** action groups work with network requests to execute the actions, the Outlook actions address the functionality of a locally installed Outlook client. This has the following consequences:

- When using Outlook, a local program is started; for example, this program must also be available, installed, and set up.

- The advantage is that the actions can be used with all mailboxes that Outlook is set up for, including Outlook 365, local Exchange Server, or any other mailbox.

- The use of Outlook Actions also offers more advanced options such as **Save messages** or **Reply to a message** for this very reason.

- If Outlook is not installed on the computer running the UI flow, one of the other action groups must be used.

- However, this also has the advantage that local computer resources can be saved and, thus, resources can be made available for other tasks.

The lightest option for emails is to use the **Retrieve email messages**, **Process email messages**, and **Send email** actions in the **Email** action group. The creation of the connection to the mail server must be defined in each of the three actions, so there are no separate actions for establishing or closing the connection. However, this only requires exactly one action, for example, to receive emails. The address of the mail server is required as well, as a valid username and password to work with the actions.

If you want to work with the actions for accessing a local Exchange Server, please note that Exchange Server up to version **2013 Service Pack 1** is supported.

Apart from the differences presented here, the actions for working with emails all work the same. The only thing to decide is whether the Outlook client can and should be used or not. This decision has to be made individually and I would like to leave it to you to experiment with further actions for emails.

We will learn which other web-based protocols and formats can be used with PAD in the next section.

Working with web protocols and formats

In the previous section, we discussed various web protocols such as LDAP and IMAP, and saw how they are supported by PAD. Next, we'll move on to probably the most common and well-known protocols FTP and HTTP, before we take a closer look at the different data formats for the web

FTP

Not so common nowadays, FTP was and still is a very efficient way to upload and download folders and files to and from a server. PAD is also able to use this protocol and access an FTP server. The following actions are provided within the FTP actions group:

- **Open / Open secure / Close connection**: These are required to establish a connection to the FTP server.

- **List directory / Change working directory**: Comparable to **DIR / CD command** in Windows command prompt to list the content of the current directory or change the current directory.

- **Create / Delete / Synchronize directory**: These actions do exactly what they are named after. We will look at the synchronization action at the end of this list.

- **Upload / Download / Delete / Rename FTP file(s)**: These are self-explanatory and simple to use.

- **Invoke FTP command**: Several FTP commands can be issued on the FTP server, such as FEAT (list extended features) or STAT (returns information about a current file transfer).

The certainly most interesting action here is the **Synchronize directory** action. Instead of comparing remote files and directories with different other actions, this one is really a big time-saver, but it can only work in one direction at a time. It offers the following action parameters:

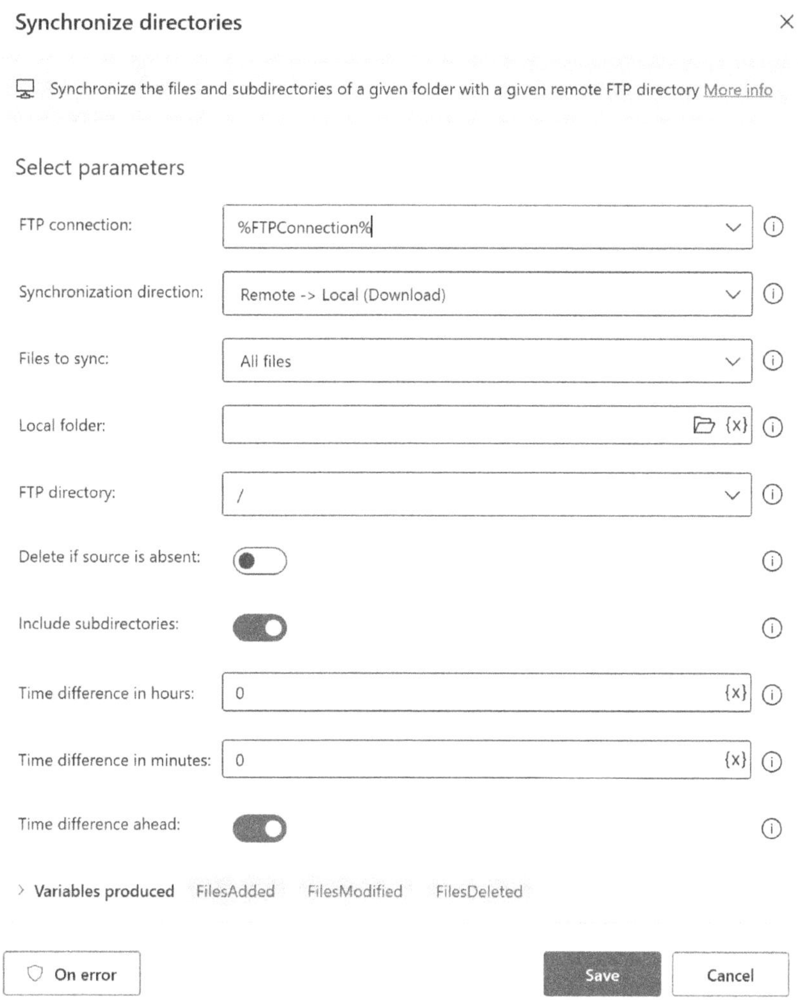

Synchronize directories   ✕

🖥 Synchronize the files and subdirectories of a given folder with a given remote FTP directory More info

Select parameters

FTP connection:	%FTPConnection% ⌄	ⓘ
Synchronization direction:	Remote -> Local (Download) ⌄	ⓘ
Files to sync:	All files ⌄	ⓘ
Local folder:	🗁 {x}	ⓘ
FTP directory:	/ ⌄	ⓘ
Delete if source is absent:	◖● ◗	ⓘ
Include subdirectories:	◖●◗	ⓘ
Time difference in hours:	0 {x}	ⓘ
Time difference in minutes:	0 {x}	ⓘ
Time difference ahead:	◖●◗	ⓘ

> **Variables produced**   FilesAdded   FilesModified   FilesDeleted

♡ On error   Save   Cancel

Figure 11.6 – Synchronize FTP directories with PAD

As you can see in the previous screenshot, this action provides all the setting options that are required for directory synchronization. The option to specify one or more filters for the files brings a lot of flexibility. The action produces three result lists, which can be used in the following:

- `FilesAdded`
- `FilesModified`
- `FilesDeleted`

While FTP was primarily designed for file exchange, the following protocol is responsible for the vast majority of internet traffic: HTTP.

HTTP

HTTP is the standard protocol on which most traffic on the internet is based. The PAD actions in this group represent API calls that PAD can perform, as well as file upload and download. Any API call works according to the request-response principle, where different methods (verbs) are available to describe the desired action. The most common methods used in such requests are the following:

- GET: This allows querying content via the URL

- POST: This is used for requests where the supplied contents are to be processed in a target resource, for example, saving a data record or other transactions

- PATCH: This represents a request for the target address to update the partial aspects transmitted

API services are very commonly used when companies want to provide their services and data controllable electronically or interact with other services. Modern web APIs thus represent the interface to the data and processes with which electronic communication can take place. While many such services use standard methods (as described previously) and are based on the **representational state transfer (REST)** principle, some services use the **Simple Object Access Protocol (SOAP)** protocol. SOAP was created to establish a standard for communication between the client and server. However, SOAP is (contrary to its name) more elaborate and complex to implement and expects a predefined structure. This and other factors have led to a tendency for new web services to be built based on the REST architecture. In PAD, of course, we have corresponding actions for both versions that can be used for this, plus another action:

- **Download from web**: Used if we want to download a specific text or a file from a web server.

- **Invoke SOAP web service**: This is the action to use if we need to communicate with a web service that is based on SOAP. This action also provides a little wizard that helps to determine the values for the different parameters.

- **Invoke web service**: This action calls a RESTful web service.

PAD manages to hide the possible complexity in web service communication. Sometimes, however, more advanced settings have to be made because, for example, certain settings, such as authentication, are expected. Therefore, the two actions for web service calls can also make advanced settings, which are stored in the **Advanced** area, such as **Connection timeout**, **User agent**, **Encoding**, and more.

Working with web data formats

When applications communicate with each other via web services, this is also done using a language. REST-based web services use JSON as a language here, which is built-in to PAD. There is, for example, an action to convert a variable into JSON and back.

> **Background information – custom object versus JSON object**
>
> Internally, PAD can handle JSON objects but prefers to use a so-called custom object. This is a more flattened structure that is easier to handle in loops and contains a list of attribute names and values. A JSON object, instead, can be of any complexity with multiple levels of nesting structures. It is possible to convert such a structure into a custom object. In this case, the custom object contains this flattened structure with nested objects as values.

Another very well-known concept is markup languages. The two best-known representatives here are **HyperText Markup Language** (HTML) and **XML**. HTML is used to represent content and web pages, and XML is for the structured storage of data or configuration information but also for communicating with web services. SOAP, as mentioned previously, communicates via XML. The basic structure of a simple XML file is depicted in the following diagram:

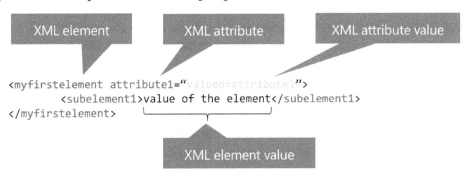

Figure 11.7 – XML structure explained

There is also a special notation to describe the way to an element or a value within an XML document. This is called `XPath` and looks like this: `/myfirstelement/subelement1`. This would give us the value of `subelement1`, which is `value of the element`. This is also referred to as an XPath query. PAD also includes actions to handle XML:

- **Read / Write XML file**: This handles the XML file that should be used in the UI flow.

- **Insert / Remove XML element**: This allows us to add a new element at a specific position of the XML document, which is provided by an XPath expression. The XML element to insert can also be a larger structure as long as there is only one root element. To remove an element, an XPath expression has to be provided.

- **Get / set XML element value**: As can be seen in the previous example, an XPath expression can also be used here to read or set the value of an element.

- **Get / set / remove XML element attribute**: These correspond to the operations that can also be performed for the XML elements, only here with the attributes.

- **Execute XPath expression**: This provides an XPath expression to select or even search for specific elements (nodes) or attributes. Please refer to the *Further reading* section to find out more on this topic.

To further complete the list of data formats, of course, a specific document format must be defined. Adobe PDF is the de facto standard for generating and exchanging documents. These are characterized by many important features that are required in the internet-based communication world:

- They can be read from almost any device and browser

- They are considered safe (do not contain viruses or malicious code)

- They cannot be modified or changed if the author does not allow it

- Their creation and consumption are free of charge

In particular, because in the context of business applications, we usually also work with documents and PDFs, PAD has a set of actions that allow working with these types of documents. These include the following:

- **Extract text / tables / images from PDF**: These actions extract content from a PDF file. The images for example can be extracted and stored in a specific location of the workstation.

- **Extract PDF file pages to new PDF file**: This action allows us to create a new PDF document from the specific pages of another document.

- **Merge PDF files**: This takes a list of PDF documents and creates one combined document.

Please note that there is no action to create a new PDF document from other content, for example, a Word document or Excel. This functionality must be provided via the respective application itself.

Last but not least, we are also able to natively compress files or folders and also uncompress a file with PAD. The **Compression** actions group contains the two actions for this. If we need to create an archive (ZIP file), we can provide the following parameters:

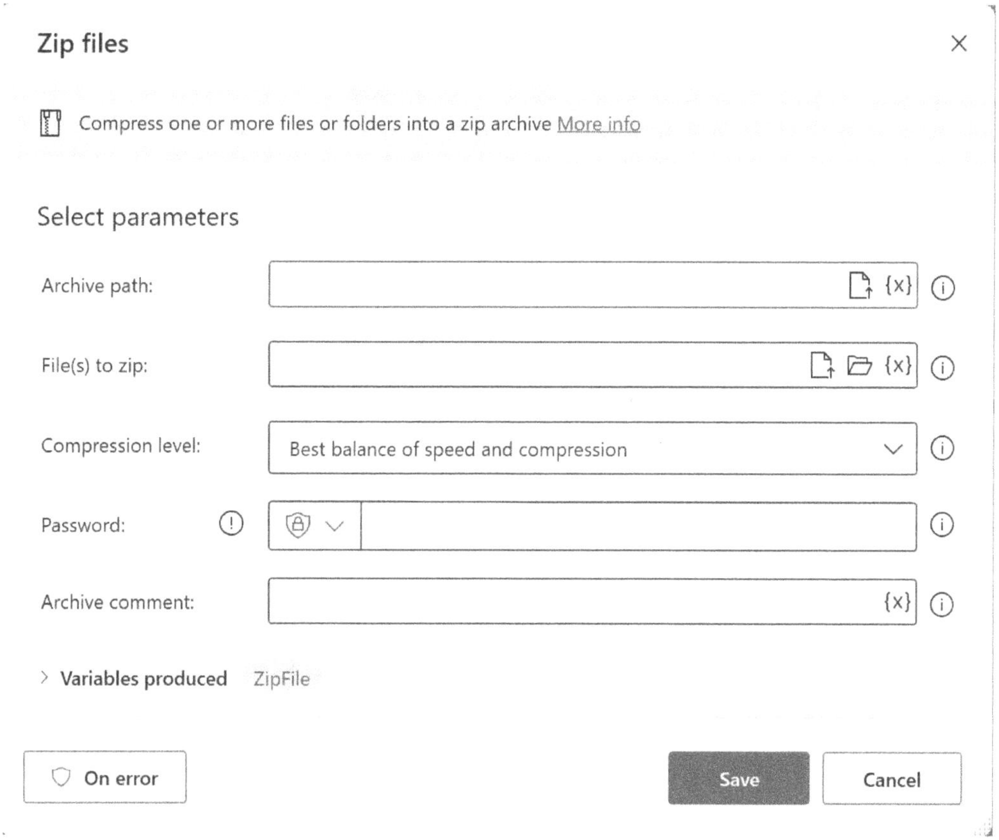

Figure 11.8 – Creating a ZIP archive in PAD

As we can see here, we can specify not only the archive and the files to be used but also the compression level and, if necessary, an archive password, which leads us directly to the next section, where we will talk about additional security mechanisms.

Using additional security features

In the security environment, we have two action groups in PAD dedicated to this issue. In the **Cryptography** group, actions are available that enable the encryption and decryption of text or files. A so-called symmetric encryption method based on the **Advanced Encryption Standard** (**AES**) algorithm is offered, which allows a maximum key length of 256 bits.

Another option in this action group is to create a hash value for a text or a file. This hash value can also be calculated based on a key that is provided. Hash values are typically used when a text or a file should be checked against changes.

These mechanisms can, of course, fulfill a basic need for security. However, care must also be taken to ensure that the key information is stored securely and is not accessible to everyone. One way to do this is to use the actions in the **CyberArk** action group. CyberArk is a provider of identity and access management security solutions. After service access is set up and a password or key is stored there, we can use the **Get password from CyberArk** PAD action to retrieve that information.

Microsoft Azure also provides a resource to store credentials in a safe way: Azure Key Vault. It can also be accessed through web services, and we will see an example of this in the next chapter.

Built-in data handling functionality

In the final section of this chapter, we will look at other PAD actions that have been used repeatedly throughout this book and the examples that have not yet been fully explained. These very useful functions include date and time functions, manipulating text, and message boxes.

Date time actions

Of course, we also need functions within PAD to work with data, to create, transform, and display it. Some of these functionalities can also be found in the action group for variables, for example, creating a random number and dealing with lists and objects. However, there are also frequent situations, especially in the business context, where it is necessary to work with date values:

- The current date should be saved for a quotation

- The due date for an invoice is to be calculated 2 weeks from now

- The age in days between two dates is to be calculated

For these purposes, the actions in the **Date Time** group can be used. Here we have the following actions:

- **Add to datetime**: This expects a value for a date and a number to add. A **Time unit** field allows us to determine whether the value to add is interpreted as seconds, minutes, hours, days, months, or years.

- **Subtract dates**: This needs two dates to subtract and produces a `Time difference` variable. The resulting number can be influenced by a third parameter, which allows us to choose a value between days, hours, minutes, and seconds.

- **Get current date and time**: This produces a variable that contains the current date and time. It is possible to use different time zones.

Actions for working with text

In addition to date and time, manipulating text values is also very important. PAD offers numerous actions for this as well, which reside in the **Text** action group. Apart from the language, you can select what exactly should be recognized:

- **Conversion actions: Convert text to number / number to text / text to datetime / datetime to text / replace**

- **Creation actions: Create random text, Recognize entities in text, and Join / Split / Append line to text**

- **Manipulation actions: Pad / Trim / Reverse / Change case / Escape text for regular expression**

- **Extraction actions: Get subtext / Crop / Parse**

The individual actions are largely self-explanatory and the Microsoft documentation contains good information on this. However, I would like to take a closer look at the following actions.

- The **Recognize entities in text** action is very helpful if there is a requirement to identify specific topics or facts in a given text and the exact position is unknown. Apart from the language, you can select what exactly should be recognized. Entries range from date and time to dimension, temperature, currency, email, URL, IP address, **globally unique identifier** (**GUID**), and more. Imagine you need to determine a delivery date or phone number from a supplier's email response. A regular expression could be used for this task, but this requires knowledge about how it works. It is simpler to use this action. It generates a result variable as a list containing the identified terms. A suitable example of this has already been presented in *Chapter 2*.

- Another useful function is **Create random text**, as shown in the following screenshot. As you can see in the screenshot, numerous parameters can be set for creating random text. This action is especially useful for creating random passwords or keys for encryption.

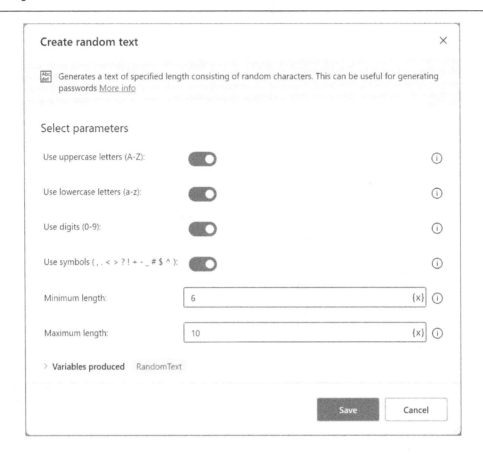

Figure 11.9 – The Create random text action properties

Last but not least, let's take a look at the actions located in the **Message boxes** group.

Interactive flows with Message Boxes

Message boxes are used many times throughout the book. These actions help make flows interactive. It is not only possible to display variables or texts in a message box. It is also possible to request user input, select files, and so on. Of course, these entries can and should be used in subsequent actions. There are dialogs for displaying messages, and inputs, selecting dates or entries from a list, and selecting a file or a folder.

It is also noteworthy that there is the option to create your own dialogue. To do this, the **Display custom form** action must be dragged onto the designer. A button is shown to open **Custom form designer**, which looks like this:

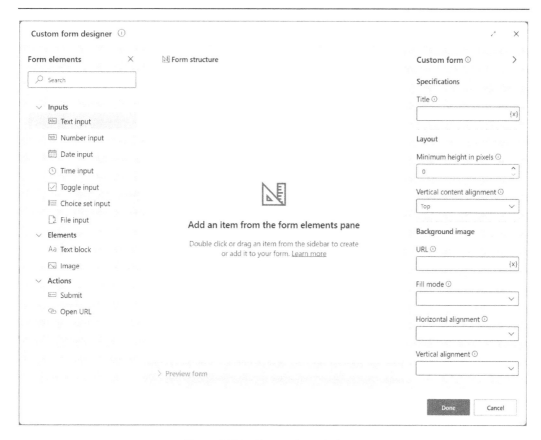

Figure 11.10 – Custom form designer

This option allows you to create dialogs with multiple inputs and outputs and can thus be adapted to specific situations. For example, for a more complex task, a dialog should be displayed at the beginning of the flow that collects all the necessary inputs and settings for the overall flow and applies them during execution. More about this, however, in the next chapter.

Summary

This chapter has shown that PAD can integrate and automate applications that do not have an API. The ability to address systems with interfaces makes PAD a complete and self-sufficient tool for automation since, typically, in automation scenarios, both legacy applications and modern applications with interfaces have to be integrated. The actions presented in this chapter allow PAD to perform all tasks that would otherwise be covered by a cloud flow, such as including a web service, transmitting and receiving data, and further processing in a UI flow. This makes PAD quite suitable as the sole tool for small to medium integration and automation tasks. However, this can also become complex in some circumstances, as it is not possible to use a dedicated connector for a web application, for example, which can hide a lot of complexity. Furthermore, there are hardly any possibilities for scaling on a local machine with PAD.

We'll take a closer look at how we can meet these challenges in the next chapter and also see an example of how the actions presented here come into play.

Further reading

- Power Automate Desktop AD actions reference: `https://learn.microsoft.com/en-us/power-automate/desktop-flows/actions-reference`

- AD: `https://en.wikipedia.org/wiki/Active_Directory`

- PAD custom objects and JSON: `https://learn.microsoft.com/en-us/previous-versions/troubleshoot/winautomation/product-documentation/best-practices/variables/custom-objects`

- XPath Syntax: `https://www.w3schools.com/xml/xpath_syntax.asp`

- Text action reference: `https://learn.microsoft.com/en-gb/power-automate/desktop-flows/actions-reference/text`

12

PAD Enterprise Best Practices

The previous chapters have focused on the detailed functionality of PAD as such. We have seen numerous **user interface** (**UI**) flow examples that have been executed interactively. However, deployment in large enterprises typically looks somewhat different. Here, UI flows are usually part of a larger concept in which issues such as security and scaling play an important role. This chapter focuses on these aspects:

- Typical requirements in enterprises, including development aspects, installation, scalability, security, governance, and compliance, and how to address them with PAD

- Microsoft best practices for hyperautomation and the Automation Kit

- Representation of an enterprise automation scenario with PAD

Chapter prerequisites

In this chapter, we will learn about the functionality of PAD, most of which is only available with the Premium RPA features. This requires a corresponding license, also described in more detail in the *Compliance* section. Microsoft offers a 30-day trial period for these features, which is suitable for trying out these functionalities.

Furthermore, several local or virtual Windows machines are typically used in such scenarios, for example, to test load balancing. For more information, see the *Installation and deployment* section.

Typical requirements in enterprises

So far, we have focused on the functionalities of PAD and looked at individual UI flows that should perform a specific task. This allowed us to look at the automation options for a certain business process and implement them with UI flows. However, in an enterprise context, many organizations today generally have additional challenges, which we will closely examine.

Hyperautomation

Automating business processes typically requires taking both legacy and new technology into account. Application landscapes at companies are often very diverse starting from cloud and on-premises systems, applications with and without APIs, existing workstations and mobile applications, applications used for mass data processing, and so on. This may also require a whole set of tools to connect and automate all these applications, and in most cases, various tools are already in use. Establishing a unified approach here typically requires the use of technologies that can handle APIs, UIs (RPA), and AI. This triad has given rise to the term **hyperautomation**, as with this toolbox, compare can be enabled to meet any automation need and thus become more efficient and productive.

Development aspects

As we already know, PAD UI flows are developed using the corresponding Windows application. We also already know that the application is always connected to a Power Platform environment. This is where the UI flows are physically stored. Access to the environment is controlled via the user. It is important to think about an environment strategy that is used in connection with Power Automate Desktop and also all other Power Platform components. In addition to this, the following aspects should be considered:

- **Use a common data model**: Use a common data model to ensure consistency and accuracy of data across all workflows. Perhaps a larger portion of the data model could be mapped directly into Dataverse, as this repository brings many features that an enterprise-wide data model requires. However, PAD also integrates and automates applications that cannot be inserted here. Therefore, it is necessary to design a conceptual overall data model with all integrated applications in the business process.

- **Use a centralized container**: To package all artifacts that are required for an automation project, it is recommended to use a centralized repository to store and manage workflows, data, and other resources. These repositories are called **solutions** in the world of Power Platform. A desktop flow is solution-aware, meaning it can be contained in a solution. The solution can then be exported and imported into another environment by using application life cycle methods.

- **Leverage application life cycle management (ALM), including a version control system as well as test and validation processes**: Typically, there are different stages where applications and solutions are developed and operated, such as *development*, *test*, and *production*. ALM technology allows transporting such solutions from one stage to another in an automated way. Use a version control system to manage changes to workflows and other resources. This area also includes the usage of a version control repository such as Git and test processes to ensure that the solution also works in subsequent stages. Read more about these topics at `https://learn.microsoft.com/en-us/power-platform/alm/`.

The following diagram illustrates best practices regarding development approaches:

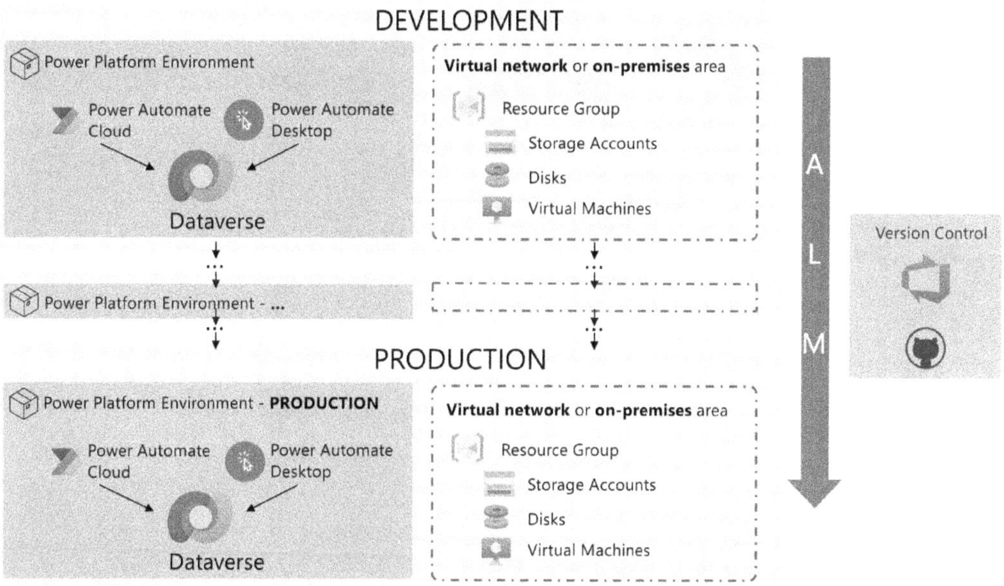

Figure 12.1 – Enterprise development approach with PAD

In the special case of PAD, we also need to consider that UI flows are running on a physical or virtual desktop. These machines need PAD installed locally and can run in the company's local network or virtually in Azure or other any other cloud provider offering **virtual machines** (**VMs**). That's why it is also important to cover this part of the solution and make sure that VMs for development, production, and all other stages are maintained. We take a look at the options we have here in the next section.

However, with all this to consider, let's also not forget that the concepts introduced in this book, such as child flows, exception handling, and more, help to create meaningful and robust UI flows.

Installation and deployment

The PAD application is a prerequisite and must be installed on the physical machine or VM used to execute large automation scenarios. Let's not forget that we will eventually need to install a legacy application on that machine. Since the manual option does not scale very well, there are also other options:

- Use systems management software such as Microsoft **System Center Configuration Manager** (**SCCM**) or Microsoft Intune to deploy and configure local machines

- Use VMs in Azure or any other cloud provider to provide virtual environments for PAD and corresponding management tools

Microsoft offers to use its **Desktop-as-a-Service (DaaS)** functionality called **Azure Virtual Desktop** to manage VMs for PAD directly within Power Platform. A starter kit has been created, containing different flows and an additional app to configure the setup (see the following screenshot):

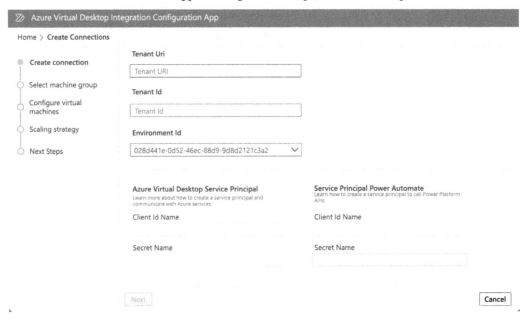

Figure 12.2 – The Starter Kit Configuration App

This option requires some setting parameters to provision appropriate machines in the background. However, this also allows adding (scale-up) or removing (scale-down) additional machines very easily, which can be very helpful in load peak periods. More on this in the next section.

> **The Starter Kit is not an official Microsoft product**
>
> It should be mentioned that the Starter Kit is not an official module in the Power Platform and should be understood as a sample implementation to achieve automatic provisioning and scaling with Microsoft technologies. All implementation artifacts are published on GitHub. Microsoft does not offer any support if there is an issue with this product, as it is not officially supported. Instead, customers are encouraged to contribute to the GitHub project and raise their issues there.

Another aspect of deployment is the deployment of the UI flows itself. When a user creates and saves a UI flow, this is stored in the Dataverse repository for the given environment. By default, however, other users cannot view or use this flow.

There are multiple ways to provide access to a UI flow to another user. The following diagram shows these options:

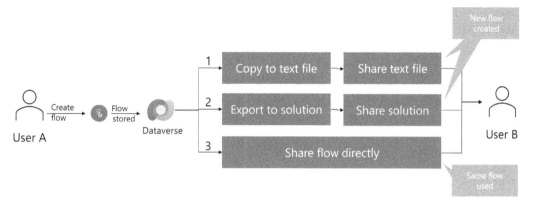

Figure 12.3 – Ways to share a flow

Please note that options **1** and **2** create a copy of the original flow. Let's look at the details.

Using a text file

The first option is to copy the flow and its actions to a text file and share this with a user. To do this, you need to take the following steps:

1. Open a UI flow.
2. Select all actions by pressing *Ctrl + A*.
3. Copy the actions to the clipboard by pressing *Ctrl + C*.
4. Open a new text document (for example, with Notepad).
5. Paste the content of the clipboard to the text file.
6. Save the text file.

We can now share this text file with another user who can open it, copy the content, and paste it into a blank UI flow designer window.

Using a solution

As mentioned before, solutions are containers that can contain various Power Platform artifacts and thus can be considered a complete business solution package. They represent a very important concept in Power Platform and are used to transport artifacts such as apps, flows, and so on from one environment to the next using ALM techniques. To create a solution, go to the **Solution** menu item in the **Power Automate** portal and click the **+ New solution** button in the menu bar, as shown in the following screenshot:

Figure 12.4 – Create a solution in Power Automate

A dialog will appear on the right-hand side, where you must enter the solution's data. The general solution concept is explained very well in the Microsoft documentation, including the publisher, versioning, and so on. For this example, the default publisher can be selected so that the solution is created. New solution components can then be created, or existing ones added to the solution. We can now use the **Add existing** button to select a desktop flow and add it to the solution (see the following screenshot):

Figure 12.5 – Add existing Desktop flow to a solution

You will be presented with a list of desktop flows that reside in this environment and you can add them to this solution. Afterward, there is an **Export solution** button, which creates a downloadable ZIP archive that can be shared with other users and other environments. Just like the previous solution, importing this solution creates a copy of the UI flow. Changes to the original flow are not applied to this copy; you will need to export/import solutions as well to update the copy of the flow in the second environment with the changes applied to the first one.

Managing and sharing a UI flow in Power Platform

The easiest way to make a UI flow (and indeed any other artifact) available to another user is to use the **Share** feature in the **Power Automate** portal. The following screenshot shows where to find this function:

Figure 12.6 – Share feature in the Power Automate portal

In the left navigation panel, go to **My flows**, and afterward, switch to the **Desktop flows** tab. You will see a list of all Desktop flows in this environment. By selecting one flow, you can press the *Share* icon in the corresponding row or the **Share** button in the top button bar to launch the sharing dialog. This dialog lets you choose another user to share the flow with. You can also specify whether the user is just able to see and execute the flow (user) or will also be allowed to change it (co-owner). If there are flows that have been shared with you, these will appear under the **Shared with me** tab. Please notice that we are not creating a copy here. This is especially important if you grant co-owner access to another user. If you want to create a copy, you could use the **Save as** button in the top button bar.

Another option would be to assign a flow to another user. This would hand over the ownership to that user. To accomplish this, we need to edit the corresponding flow in Dataverse in the environment where this UI flow is stored. For simplicity, we can think of Dataverse as a relational database with various business process tables. UI flows are stored in a table called `Process`. The documentation (refer to *Manage Desktop flows* in the *Further reading* section) describes well which steps have to be taken for this. Reassigning a UI flow to another owner can be useful if the creator of the flow should not be the executing user, for example, because this user needs special permissions or already owns and should execute other flows.

Please note that it is also possible to execute administration operations on UI flows using web APIs. Just like any other artifact in Dataverse, it is possible to list available desktop flows, get schema information and status, get flow output, and, of course, trigger flows or cancel UI flow runs. This means that UI flows can be called not only via the corresponding connector in Power Automate but in principle, also by any program or script. It offers a whole range of new possibilities regarding integration and could lead to a situation where a whole series of UI flows is to be executed in parallel or at a high frequency. That's why we discuss how to create a scalable PAD environment in the following section.

Scalability

For an enterprise-ready deployment, the scalability of the processing platform is an important criterion. This means that it must be possible to distribute any peak loads to several working nodes and, if necessary, implement a queuing concept to ensure that requests are processed in an orderly and reliable manner. This can be achieved by one or a few individual machines or by load-balance management with corresponding machine management, which we will now take a closer look at. For both scenarios, we assume that the machines are registered within Power Platform so they can be included in a cloud flow and that the legacy applications are installed on the corresponding machines. Please refer to *Chapter 9* for more details.

A single-machine approach

Sufficient capacity in terms of **random-access memory** (**RAM**) and processor time could result in better-equipped workstations on which PAD is installed and used for processing. In fact, by using Windows Server 2016, 2019, or 2022 and running flows unattended, it is possible to have multiple flows running simultaneously on the same machine, thus reducing infrastructure costs. This scenario can be illustrated like this:

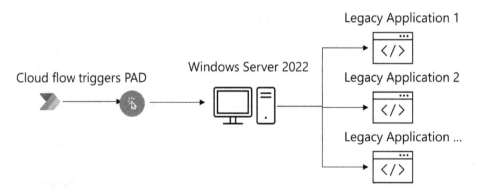

Figure 12.7 – Scalability with a single machine concept

However, this would also require adjustments to the UI flows as they would need to be run by different users to create different user sessions on the server. But this scenario runs the risk of creating a single point of error. In other words, if the machine goes down, no single UI flow will run.

Machine groups and flow queues

A more dynamic approach is the use of machine groups and a load-balancing concept provided by Power Platform directly. The concept is depicted in the following diagram:

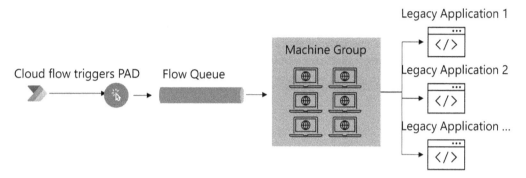

Figure 12.8 – Machine groups and flow queues

In this setup, a machine for UI flow execution is registered in Power Platform and also inserted in a machine group. In the corresponding cloud flow action for PAD, we can select this machine group as the execution environment, and everything else is handled by the platform. If several flows need to be executed simultaneously or in quick succession, the requests are first pooled in a queue. The platform then decides which machine to send an execution request to and distributes the request load evenly. If a machine should fail here, this initially represents a lower capacity on the one hand but does not lead to the complete collapse of the overall processing. On the other hand, it is very easy to add more machines to the pool to increase the total capacity without interrupting the process. The automated deployment of workstations with PAD, along with silent registration and silent join of a machine group, allow this to carry out.

Microsoft also offers the option to use VMs for UI flow execution directly in the platform, which can be provisioned by using the Starter Kit (also refer to the *Installation and deployment* section). With that option, it would also be possible to define a scaling strategy, which allows measuring the load and provisioning or deprovisioning additional machines.

Security

Among the various requirements for security in automation scenarios, we will take a closer look at two aspects in this section that are important in the context of PAD and for enterprise use: **data loss prevention** (**DLP**) policies and secure credential storage.

Protecting data with policies

In every integration and automation scenario, data plays a central role. At the same time, it is precisely this data that is also a very valuable asset worthy of protection, so it must be ensured in every case that no data leaks out unintentionally. Especially in environments such as Power Platform, where essentially every employee can help optimize processes through low-code applications, policies must be able to be applied to prevent unwanted data flow. Data access and provision (sending to a receiver) are done using connectors, which we already learned about in *Chapter 9*. We can now use the DLP policies, with which it is possible to define which connectors can be used with each other in a flow and which cannot. This can prevent, for example, a connector for an internal system being used in the same flow as a connector for the external provision of data and thus inadvertently leaking data to the outside. DLP policies are defined in **Power Platform admin center**, as depicted in the following screenshot:

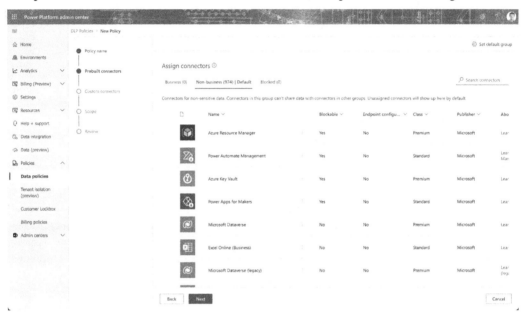

Figure 12.9 – Definition of DLP policies

Note on DLP policies

At the time of writing, this feature is in preview, and the use of these capabilities must be confirmed by an opt-in (announcement to Microsoft). Subsequently, the action groups and also individual actions can be blocked or restricted so that UI flows are disabled if a DLP policy is violated in the flow (prohibited connector or unauthorized combination with another connector).

When the feature comes into general availability, it will only be available for managed environments for which additional security mechanisms exist beyond that. The *Further reading* section contains a link to get more information about managed environments.

Apart from the actual data, there is also other information worth protecting, such as access data and passwords. We will look at the possibilities for this in the next section.

Secure credential storage

As we have already seen when including a UI flow, the credentials for the Windows computer must also be entered so that a login can take place and the call to the corresponding UI flow can be performed.

Here, the Desktop Flow action used in a Cloud Flow follows the same concept as all other connectors. The login information is not stored directly in the flow. There is a separate storage mechanism for connections, which is encrypted. The credentials are only accessible to the user who created them, even if the UI flow is shared with another user.

But what if we need to work with credentials within a UI flow or any other component? To store passwords or other sensitive information securely, there is **Azure Key Vault** (**AKV**), a resource in Azure that allows storing access keys, passwords, and certificates. AKV as a secure repository for sensitive information is a core concept and therefore numerous **software development kits** (**SDKs**) and web APIs are available to work with it. We can also access AKV within a UI flow, for example, to retrieve and use access information for an application or another system. This process is exemplified in the following diagram:

Figure 12.10 – Using AKV in UI flows

As you can see, access to AKV using the corresponding web APIs requires additional tasks because to access this resource, the user first needs to authenticate against the platform, retrieve the access token, and then use this token in subsequent requests for the resources. Microsoft has announced that there will be a PAD action for accessing AKV, but this has not been released yet. The *Further reading* section contains the link to the announcement for this. It is strongly recommended to use this service for the storage of sensitive information. Now let's take a look at additional considerations for governance and compliance.

Governance and compliance

In Microsoft's documentation, there is a separate article on the topic of governance, which describes in detail the settings for the Windows operating system (registry settings) to restrict and monitor the functionality of PAD. We want to take the topic of governance a bit further here and look at what options are available to monitor UI flows and thus monitor the overall operations. Furthermore, on the topic of compliance, we will take a look at licensing.

Monitoring flow execution

Two portals can be used to monitor the execution of flows: the **Power Automate** portal and **Power Automate admin center**.

The **Power Automate** portal provides a more operational monitoring capability. These functions are addressed to those who want to develop the flows, test them, and also monitor their operational functionality. The first view provides various charts, aggregated views, and filtering options (see the following screenshot):

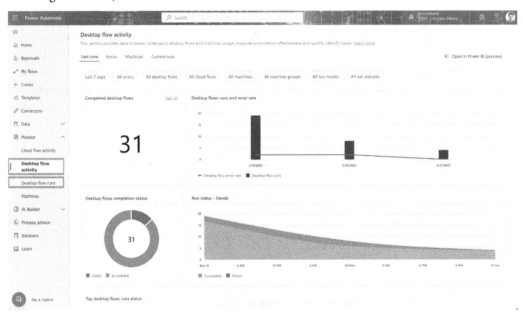

Figure 12.11 – Monitoring PAD flows

The second view in the **Power Automate** portal provides a section for monitoring the real-time execution of Power Automate Cloud and UI flows. This shows each execution of the flows and their status: **Queued**, **Running**, **Succeeded**, or **Failed**. This makes it very easy to monitor which flows are routed to which machine and which flows fail.

The reports in **Power Platform admin center** are aimed more at administrators who monitor the workload in the context of the environments. To do this, we need to go to Power Platform admin center and go to the **Analytics | Power Automate** section (see the following screenshot):

Figure 12.12 – Flow monitoring in Power Platform admin center

There are different tabs available to provide information about the runs, usage, number of flows created and shared, and analysis of the connectors being used in flows. Governance requirements about the monitoring of Power Automate can thus be addressed very well. Another part is licensing, which we will look at in the following section.

Licensing

In the course of the book, we have already learned about many functionalities of PAD, most of which are available with the free version. This is especially true for single use on a workstation, where flows are started manually and run interactively. Requirements such as scaling, parallel execution, centralized management, and rolling out flows are more typical for use in large enterprises and require the appropriate licensing. There are several license types available for this purpose, which then unlock a number of features and benefits and include the following features:

- Automatic scheduling/triggers from Cloud flows
- Flow triggering from the desktop shortcut or via a URL
- Access to multiple environments and sharing and collaboration
- Centralized management, monitoring, and Desktop flow analytics

It, therefore, makes sense to provide an appropriate number of licenses for the organization to support the automation projects in the best possible way. The consumption of licenses can be viewed very easily in the admin center of Microsoft 365, where all other licenses are also listed. In addition, Microsoft has announced that it will also offer license reporting directly in Power Platform admin content. The corresponding menu item already exists but is currently without content.

The previous sections have shown that a number of tools exist that qualify PAD for use in large enterprises. In addition, some concepts build on this to help companies adapt low-code technologies and shape a transformation toward higher levels of automation. The following section describes this concept within Microsoft.

Microsoft best practices for hyperautomation

The extent to which companies are able to connect their processes through automation technology can be determined by a maturity model. Microsoft suggests the following Power Platform automation maturity model, shown in the following diagram:

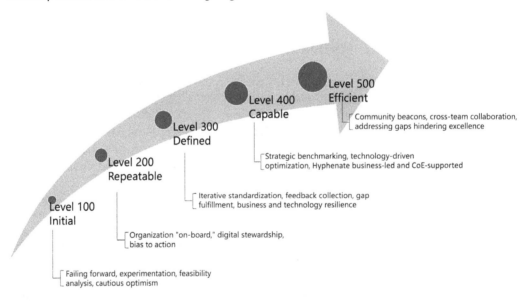

Figure 12.13 – The Power Platform maturity model

This model can help determine the current state of automation capabilities and, from there, take the next steps, if necessary, to achieve a better level of automation and thus increase the company's efficiency and productivity. In general, the idea here is not to look at individual applications but to take a holistic view of automation practices across an enterprise to embed them in an overall concept. For each individual automation project, so-called *holistic enterprise automation techniques* (HEAT) techniques can then be used to help design and roll out the solution in the context of the organization and manage the overall life cycle of the solution.

Another important building block is to build a team that addresses the various aspects of an automation solution and includes or builds experts to look at the solution from any facet, such as business content, ALM, infrastructure, security, and governance, as well as operational reporting and analytics. Another term for such a team is **Center of Excellence (CoE)**, and in relation to automation projects, **Automation Center of Excellence**.

In the meantime, many companies have also implemented these concepts in tools and processes. Microsoft has embedded this information into a so-called Automation Kit and, thus, a collection of applications and flows based on Power Platform. This kit can be installed and configured as a solution in an environment. The basic idea is to use the artifacts and apps installed there to support an automation project from start to finish, including the following aspects:

- Definition of goals
- Review and release of the project
- Development and deployment
- Operational use and monitoring

This also supports different enterprise roles such as CVP, process owner, end user, and others. The following diagram illustrates the life cycle of such a project:

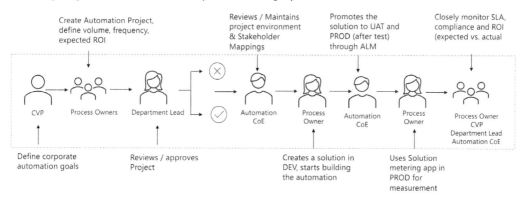

Figure 12.14 – Automation project life cycle

The fact that the kit can also be customized and that not all functions necessarily have to be used makes it easy to get started and shows what is possible with this technology. It is thus definitely worth a look and can bring added value and acceleration to automation projects.

Enterprise automation scenarios with PAD

We have already encountered many examples in this book, sometimes applicable to enterprises and sometimes to citizen scenarios. As we have also seen, PAD is a natural part of Power Platform and is therefore suitable for all applications that involve connecting locally installed applications. Nevertheless, care should be taken to ensure that the overall architecture also fits the requirements and the circumstances. Furthermore, there is usually always more than one way to implement a requirement. In our example, we want to map the following scenario:

- A company has external vendors who sell specific products or services to the company's customers.

- For each customer, sales opportunities exist in an internal CRM system represented by Dynamics 365 (**Software-as-a-Service (SaaS)**), which represents the potential business. The internal system is not accessible to external sellers.

- External vendors are managed by internal staff, each of whom has a group of vendors assigned to them.

- The external sellers write an email to their team manager (internal employee) when a customer has agreed to the cooperation and wants to order.

- Automated processing will be created in which email responses from external vendors will be processed. For the customers reported in it, all sales opportunities should be set to **Won**. A master data record must also be created for the customers in the invoicing program (Contoso Invoicing) and also in the ERP system (SQL Server).

- The invoice program is a legacy application and runs on a Windows computer. Because the planned automation can lead to load peaks when several team leaders start processing at the same time and, if necessary, several sales opportunities per customer have to be processed, it must be ensured that there is a scalable infrastructure for connecting the invoice program in which no processing requests are lost.

- Processing should be able to be started on demand by a team leader, preferably using a shortcut on the desktop.

- The CRM system is implemented with Dynamics 365, and Power Platform with Power Automate is also in use.

- The internal invoicing system is represented by Contoso Invoicing (a desktop application), and the ERP system is accessible via a database (SQL Server).

At this point, we probably already have enough information to sketch an initial idea for the overall process. This could look as follows:

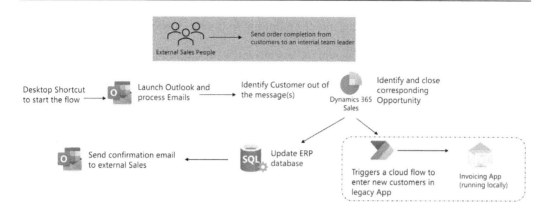

Figure 12.15 – Automation project scenario

Now, as always, there are several ways to implement this requirement, but there is a lot to be said for using a UI flow for this:

- The processing should be started individually by a team leader, which can be enabled via a shortcut on the desktop thanks to the advanced features.

- Outlook actions enable us to retrieve and process emails.

- We can extract the relevant information by using the text actions from *Chapter 11* (**Parse text** or **Recognize entities in text**). Of course, it would be beneficial if the email were standardized, making it easier to identify the customer ID.

- The next step would be to call the Dynamics 365 web API, retrieve all opportunities for the customer, and close these. An HTTP action will help us here to accomplish this action (**Invoke web service**).

- From a design perspective, we can outsource the processing for the invoice program to a separate cloud flow, which, in turn, includes the integration of the invoice program using the Desktop Flow connector. For the invoice application, we can then use the concepts for scaling that we learned about in this chapter.

- The ERP database will be updated by using the database actions, which we learned about in *Chapter 11*.

So, we can divide the overall processing into the following sections as subflows:

- `GetEmailMessages`: This should retrieve all emails and create a list of customers we need to feed into our systems.

- `CloseOpportunities`: For each customer, this subflow queries all open opportunities in Dynamics 365 and closes them. This will automatically trigger a cloud flow that listens to the **Opportunity Close** event and starts the processing with the invoicing app.

- `UpdateERP`: This will also take the customer information and create new customer records in the database.

- Subflow `NotifySeller`: This will send out a confirmation to the external seller.

You will notice during the projects that in many places, already known concepts can be reused, as in this case, we have already covered the use of Outlook, information extraction with text actions, as well as database queries in this book. So let's take a look at a topic that hasn't been looked at in detail yet: using the **Invoke web service** action to integrate Dynamics 365 from our UI flow.

Calling web services within PAD

Another very powerful tool within PAD is the **Invoke web service** action. Almost all applications today provide web services as interfaces to trigger processes or work with data, and so does Dynamics 365. However, in an enterprise environment, these web services are usually protected from unauthorized access, and some form of login must be performed. Here, it all depends on the service, as there are different options for authentication and authorization. In the case of Dynamics 365 and generally all resources in the Microsoft environment, we use login via a so-called service principal, for example, a login via an instance registered in Azure **Active Directory** (**AD**) with a corresponding password (client secret). We already learned about this type of login in *Chapter 9*, where we worked with Azure VMs. Microsoft and most other providers use the OAuth login procedure for this. The goal of this processing is to do the following:

- Log in to Dynamics 365 by providing a tenant ID, client ID, and a client secret and obtain an access token, which can be used to access subsequent services

- Query Dynamics 365 with the given customer and retrieve all open opportunities

- Close each opportunity by calling the corresponding Web Service action in Dynamics 365

The following screenshot shows this subflow:

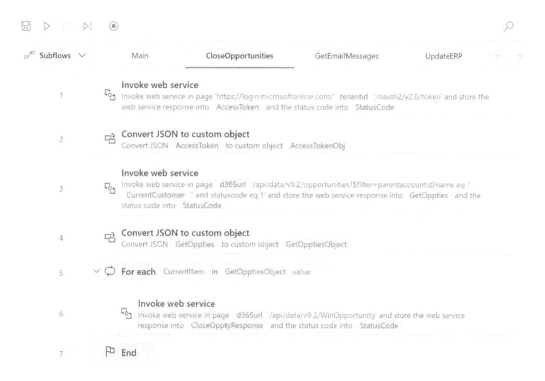

Figure 12.16 – Calling Dynamics 365 web services

This flow does the following:

1. Calls the authentication URL for Microsoft services with the required parameters to retrieve an access token. This is stored in the corresponding variable and is converted into a custom object to use in subsequent actions (lines **1** and **2**).

2. Calls the Dynamics 365 web service and a query parameter to retrieve all opportunity records for the given account name, which has been set by the previous subflow. The result of this is a list of opportunities for that customer that we need to close. The list is also converted to a custom object (lines **3** and **4**).

3. In the loop, call a specific Dynamics 365 action to close an opportunity (lines **5** to 7).

Please note that we are calling a SaaS service from a UI flow here. It would otherwise be very inconvenient to use a cloud flow for this in our current situation. The communication is very fast and stable and is a real productivity gain, as these records would otherwise have to be closed manually, which would definitely take much more time.

However, as mentioned earlier, it could be that multiple team leaders start this process at the same time. It must therefore be ensured that possible load peaks are absorbed. We will take a look at this setup in the following section.

Providing scalability and reliability with machine groups and queues

The execution of desktop flows is usually slower than that of cloud flows. There are the following reasons for this:

- A connection must be established to a remote machine
- The legacy application may need to be started
- The actual processing, calling different screens, and so on must take place

All these things take their time, during which a cloud flow may be blocked from execution. And during this process, no other flow would be able to start another process on this machine.

The queuing and scaling options in Power Platform can address these issues. The execution of desktop flows is managed by queues and can be distributed to multiple machines within a machine group.

This concept is described in detail in this chapter. For this setup here, I created two Windows 11 machines equipped with PAD and in the same machine group.

If the Desktop Flow described here is now executed on another machine, this will result in several opportunities being marked as **Won** very quickly. This happens so quickly that individual processing cannot even be started. Nevertheless, the request is recorded by the queuing, and the platform controls to which machine the next execution is sent (refer to the *Machine groups and flow queue* section in this chapter). The following screenshot shows the **Run queue** tab for a machine group that has queued some flows:

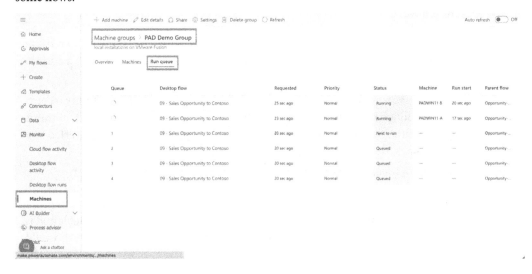

Figure 12.17 – Run queue in Power Automate

Summary

The concepts presented in this chapter build on the full range of Power Platform tools to address all automation requirements. PAD is a building block in these concepts and fits in seamlessly. The use of Power Platform also has the advantage that the general requirements for all tools, such as monitoring and management, as well as compliance with legal requirements, can be carried out at a central location.

PAD is constantly being extended and improved and already offers numerous features that enable its use in large companies and support automation projects, as we have seen in this chapter.

It is expected that deeper integration into Power Platform will occur, building on the concepts and capabilities presented in this book. This provides a very solid foundation for further automation projects based on Power Automate Desktop.

Further reading

- *Power Platform automation maturity model overview*: https://learn.microsoft.com/en-us/power-automate/guidance/automation-coe/automation-maturity-model-overview

- *HEAT*: https://learn.microsoft.com/en-us/power-automate/guidance/automation-coe/heat

- *Manage Power Automate for Desktop on Windows* whitepaper: https://learn.microsoft.com/en-us/power-automate/guidance/automation-coe/manage-pad-on-windows

- *Work with desktop flows using APIs*: https://learn.microsoft.com/en-gb/power-automate/developer/desktop-flow-public-apis

- Manage Desktop flows: https://learn.microsoft.com/en-us/power-automate/desktop-flows/manage#share-desktop-flows

- *Introduction to the Power Automate Azure Virtual Desktop integration starter kit*: https://learn.microsoft.com/en-us/power-automate/desktop-flows/avd-overview

- *Data loss prevention (DLP) policies (preview)*: https://learn.microsoft.com/en-us/power-automate/desktop-flows/data-loss-prevention

- *Managed Environments overview*: https://learn.microsoft.com/en-us/power-platform/admin/managed-environment-overview

- Administering a low-code intelligent automation platform whitepaper: https://learn.microsoft.com/en-us/power-automate/guidance/automation-coe/automation-admin-gov

- *Governance in Power Automate*: https://learn.microsoft.com/en-us/power-automate/desktop-flows/governance

- *Use credentials from Azure Key Vault in desktop flow connections*: https://learn.microsoft.com/en-us/power-platform-release-plan/2022wave2/power-automate/planned-features

- *Premium RPA features*: https://learn.microsoft.com/en-us/power-automate/desktop-flows/premium-features

- *Introducing the Automation Kit for Power Platform*: https://powerautomate.microsoft.com/en-us/blog/introducing-the-automation-kit-for-power-platform/

Index

www.packtpub.com

Subscribe to our online digital library for full access to over 7,000 books and videos, as well as industry leading tools to help you plan your personal development and advance your career. For more information, please visit our website.

Why subscribe?

- Spend less time learning and more time coding with practical eBooks and Videos from over 4,000 industry professionals

- Improve your learning with Skill Plans built especially for you

- Get a free eBook or video every month

- Fully searchable for easy access to vital information

- Copy and paste, print, and bookmark content

Did you know that Packt offers eBook versions of every book published, with PDF and ePub files available? You can upgrade to the eBook version at packtpub.com and as a print book customer, you are entitled to a discount on the eBook copy. Get in touch with us at customercare@packtpub.com for more details.

At www.packtpub.com, you can also read a collection of free technical articles, sign up for a range of free newsletters, and receive exclusive discounts and offers on Packt books and eBooks.

Other Books You May Enjoy

If you enjoyed this book, you may be interested in these other books by Packt:

Workflow Automation with Microsoft Power Automate

Aaron Guilmette

ISBN: 9781803237671

- Learn the basic building blocks of Power Automate capabilities
- Explore connectors in Power Automate to automate email workflows
- Discover how to make a flow for copying files between cloud services
- Configure Power Automate Desktop flows for your business needs
- Build on examples to create complex database and approval flows
- Connect common business applications like Outlook, Forms, and Teams
- Learn the introductory concepts for robotic process automation
- Discover how to use AI sentiment analysis

Packt is searching for authors like you

If you're interested in becoming an author for Packt, please visit `authors.packtpub.com` and apply today. We have worked with thousands of developers and tech professionals, just like you, to help them share their insight with the global tech community. You can make a general application, apply for a specific hot topic that we are recruiting an author for, or submit your own idea.

Share Your Thoughts

Now you've finished *Democratizing RPA with Power Automate Desktop*, we'd love to hear your thoughts! Scan the QR code below to go straight to the Amazon review page for this book and share your feedback or leave a review on the site that you purchased it from.

`https://packt.link/r/1803245948`

Your review is important to us and the tech community and will help us make sure we're delivering excellent quality content.

Download a free PDF copy of this book

Thanks for purchasing this book!

Do you like to read on the go but are unable to carry your print books everywhere? Is your eBook purchase not compatible with the device of your choice?

Don't worry, now with every Packt book you get a DRM-free PDF version of that book at no cost.

Read anywhere, any place, on any device. Search, copy, and paste code from your favorite technical books directly into your application.

The perks don't stop there, you can get exclusive access to discounts, newsletters, and great free content in your inbox daily

Follow these simple steps to get the benefits:

1. Scan the QR code or visit the link below

https://packt.link/free-ebook/9781803245942

2. Submit your proof of purchase
3. That's it! We'll send your free PDF and other benefits to your email directly